James Thomson Bottomley

Four figure mathematical tables

Comprising logarithmic and trigonometrical tables

James Thomson Bottomley

Four figure mathematical tables
Comprising logarithmic and trigonometrical tables

ISBN/EAN: 9783337131371

Printed in Europe, USA, Canada, Australia, Japan

Cover: Foto ©berggeist007 / pixelio.de

More available books at **www.hansebooks.com**

FOUR FIGURE
MATHEMATICAL TABLES.

MACMILLAN AND CO., Limited
LONDON · BOMBAY · CALCUTTA
MELBOURNE

THE MACMILLAN COMPANY
NEW YORK · BOSTON · CHICAGO
ATLANTA · SAN FRANCISCO

THE MACMILLAN CO. OF CANADA, Ltd.
TORONTO

FOUR FIGURE

MATHEMATICAL TABLES:

COMPRISING LOGARITHMIC AND TRIGONOMETRICAL TABLES, AND
TABLES OF SQUARES, SQUARE ROOTS, AND RECIPROCALS.

BY

J. T. BOTTOMLEY, M.A., LL.D., D.Sc.,

F.R.S., F.R.S.E., F.C.S.,

LATE LECTURER IN NATURAL PHILOSOPHY IN THE UNIVERSITY OF GLASGOW.

MACMILLAN AND CO., LIMITED

ST. MARTIN'S STREET, LONDON

1910

First Edition 1887.
Reprinted 1890, 1893, 1894, 1896, 1897, 1899 (twice), 1900,
1901, 1902, 1903 (twice).
With additions 1904, 1905 (twice), 1907, 1908, 1909, 1910.

GLASGOW: PRINTED AT THE UNIVERSITY PRESS
BY ROBERT MACLEHOSE AND CO. LTD.

EXPLANATIONS AND RULES

USE OF THE ACCOMPANYING TABLES.

THE logarithm of a number consists in general of two parts, an integer part and a decimal. The integer part is called the *Index* or *Characteristic*; the decimal part is called the *Mantissa*.

RULE I. The Index of the logarithm of a number greater than unity is the number which is less by one than the number of digits in the integral part of the given number.

Thus, the index of the logarithm of   47320   is 4.

                                      473·2   is 2.

                                      4·732 is 0.

RULE II. The Index of the logarithm of a number less than unity, and reduced to the form of a decimal fraction, is negative, and is a higher number by one than the number of zeros that follow the decimal point of the given number.

Thus, the index of the logarithm of   ·4732   is − 1

                                     ·004732 is − 3

To denote that the index is negative the sign minus is often written above it; thus $\bar{1}, \bar{3}$.

RULE III. To find the mantissa of the logarithm of a given number consisting of four figures.—Find the *first two* figures in the left hand column of the table. Pass along the corresponding horizontal line and take the number in the vertical column headed by the *third* figure. To this number *add* the number found in the difference columns under the *fourth* figure of the given number. The sum with a decimal point prefixed is the required *mantissa*.

Example. Find the mantissas corresponding to the sequences of figures 4732 and 6985.

$$
\begin{array}{llll}
473 & \cdot \quad \cdot \quad \cdot \quad \cdot \quad \cdot & 6749 \\
& \text{2 from dif. col.} \quad \cdot \quad \cdot & 2 \\
\hline
& & \cdot 6751 \\
698 & \cdot \quad \cdot \quad \cdot \quad \cdot \quad \cdot & 8439 \\
& \text{5 from dif. col.} \quad \cdot \quad \cdot & 3 \\
\hline
& & \cdot 8442
\end{array}
$$

RULE IV. To find the logarithm of a given number consisting of four figures.—Find the mantissa corresponding to the given four figures, and to it prefix the proper index. The number thus obtained is the required logarithm.

Examples.

$$
\begin{array}{llll}
\log 47320 & \cdot \quad \cdot \quad \cdot & \text{is } 4\cdot6751 \\
\log 47\cdot32 & \cdot \quad \cdot \quad \cdot & \text{is } 1\cdot6751 \\
\log 6\cdot985 & \cdot \quad \cdot \quad \cdot & \text{is } 0\cdot8442 \\
\log 0\cdot006985 & \cdot \quad \cdot \quad \cdot & \text{is } \bar{3}\cdot8442
\end{array}
$$

Note.—A logarithm whose index is negative really consists of a positive mantissa with a negative index algebraically added to it. Thus :— $\bar{1}\cdot8442 = +0\cdot8442 - 1$. It is important to bear this in mind in numerical operations on logarithms. For example, in taking the square root of $0\cdot6985$, the logarithm of that number is divided by 2, and in taking the cube by 3. The simplest way of doing this is as follows :—

$$\tfrac{1}{2}(\cdot8442 - 1) = \tfrac{1}{2}(1\cdot8442 - 2) = \cdot9221 - 1$$
$$\tfrac{1}{3}(\cdot8442 - 1) = \tfrac{1}{3}(2\cdot8442 - 3) = \cdot9481 - 1$$

RULE V. To find the anti-logarithm of a given logarithm, *i.e.*, the number corresponding to the given logarithm.—Find in the table of anti-logarithms, proceeding as in Rule III., the sequence of figures corresponding to the *mantissa* of the given logarithm. To these figures place a decimal point, in the position indicated by the index of the given logarithm, prefixing or affixing zeros, if necessary. (See Rules I and II.) The number thus obtained is that required.

Examples. Given the logarithm $2\cdot7834$ find the anti-logarithm.

$$
\begin{array}{lll}
\cdot783 & \cdot \quad \cdot \quad \cdot \quad \cdot \quad \cdot & 6067 \\
& \text{4 from dif. col.} \quad \cdot \quad \cdot & 6 \\
\hline
& & 6073
\end{array}
$$

Hence the number whose logarithm is $2\cdot7834$ is $607\cdot3$.

The number corresponding to the logarithm 6·7834 is 6073000 ; that corresponding to $\bar{4}$·7834 is ·0006073.

Note.—The use of Rules I. and II., which are commonly given for the purposes of finding the index and of placing the decimal point in an anti-logarithm, may be dispensed with altogether if the principle on which these rules are founded is kept in view; and in reality the principle is more simple than the rules and easier to remember. The logarithm, to the base 10, of any number greater than 1 and less than 10 is a positive proper fraction, and is given in the tables as a decimal without whole number. On the other hand the anti-logarithm of a decimal without whole number is a number greater than 1 and less than 10.

Thus log 7·32 = 0·8645 ; and the logarithm 0·6931 corresponds to the number 4·933.

Any number such as 7320, or ·000732 is derived from 7·32 by multiplying or dividing by a power of 10 ; and the corresponding change in the logarithm is made by adding or subtracting the index of that power of 10.

Thus $7320 = 7·32 \times 10^3$; log. $7320 = ·8645 + 3$

 $·000732 = 7·32 \times 10^{-4}$; log. $·000732 = ·8645 - 4$

In the same way since ·6931 as a logarithm corresponds to 4·933, it follows that 2·6931, or ·6931 + 2, corresponds to $4·933 \times 10^2$; and $\bar{3}$·6931, or ·6931 − 3, corresponds to $4·933 \times 10^{-3}$.

RULE VI. Given any angle less than 90° to find its natural sine, cosine, tangent, etc., or its value in radian measure.*—Find the degrees in the left hand column of the proper table. Pass along the corresponding horizontal line, and take out the number in the vertical column headed by the number of minutes lower than, and nearest to, the given number of minutes. Take the difference between the number of minutes given and the number of minutes just found, and from the difference columns find in the same horizontal line the corresponding correction. This correction is *additive* in the cases of the sine, tangent, secant, and radian measure. In the cases of cosine, cotangent, and cosecant it is *subtractive.*

Note.—It will be observed that the main division of the degree in the trigonometrical tables is into parts of 6′ each. This corresponds to *decimals* of the degree. Thus, 12°18′ = 12°·3.

Note.—In the tables of natural sines and cosines the decimal points are omitted. In the other tables the decimal points and the whole numbers which precede them are omitted in all the columns except

* Formerly called "circular measure."

that headed o′ ; and excepting also the case of a few numbers at the extremities of the tables, where the variation of the trigonometrical function is extremely rapid. At the extremities of some of the tables differences are not given, as the variation of the function is so rapid as to make the differences unserviceable.

Examples. Find the sine and cosine of 18°27′, and the tangent and secant of 58°44′.

From table of sines	18°24′	·	-	-	·3156
	3′	·	-	-	8
sin 18°27′	·	·	·	·3164	
From table of cosines	18°24′	·	·	·	·9489
	3′	·	·	·	3
cos 18°27′	·	·	·	·9486	
From table of tangents	58°42′	·	·	·	1·6447
	2′	·	·	·	21
tan 58°44′	·	-	-	1·6468	
From table of secants	58°42′	·	·	·	1·9249
	2′	·	·	·	18
sec 58°44′	·	-	-	1·9267	

RULE VII. To find the logarithmic sine, tangent, cosine, cotangent, secant, or cosecant of an angle less than 90°. Proceed as in Rule VI., using the proper table.

Note.—The sines of all angles, and the tangents of angles less than 45°, being less than unity, the logarithms of these sines and tangents are preceded by a negative index. In order to avoid the writing of these negative indices the number 10 is added to the *real* value of the log. sin. log. tan. etc., and the number so found is entered in the tables. In all calculations this must be borne in mind, and allowance must be made.

RULE VIII. To find the angle in degrees and minutes, or in degrees and decimals of a degree, corresponding to any given natural or logarithmic sine, cosine, tangent, etc. Find in the proper table the number nearest to that given, interpolating, if necessary, by means of the difference columns ; and by reversing the process of Rules VI. and VII. obtain the corresponding number of degrees and minutes, or degrees and decimals of a degree.

The preceding explanations are easily applicable to the remaining tables of squares, square roots, and reciprocals. With regard to the tables of squares and square roots, it is to be noticed that while the square of such a number as 528 is found from the square of 5·28 simply by multiplying by a power of 10, a similar relation does *not* hold always in the case of the square root. It is necessary, therefore, to have two tables of square roots—one extending from 1 or 100 to 9.99 or 999, and the other from 10 or 1000 to 99.99 or 9999.*

RULE IX. To find the Neperian or hyperbolic logarithm of a number.—If the number be greater than 1 and less than 10 its Neperian logarithm is found directly from the proper table in the manner explained in Rule III. If the number is greater than 10 or less than 1, it may always be expressed as the product of two factors, of which one is a power of 10, and the other a number greater than 1 and less than 10; the latter being simply the original series of figures with the decimal point suitably moved. The sum of the Neperian logarithms of these two factors is the Neperian logarithm of the given number. A table of Neperian logarithms of powers of 10 is given on pp. 54, 55. *Examples.*

Find the Neperian logarithms of 3 241, 324·1, and ·0003241.

$$(1) \quad \log_e 3\text{·}241 \quad - \quad - \quad \text{is } 1\text{·}1759$$

$$(2) \quad \log_e 324\text{·}1 = \log_e 3\text{·}241 \times 10^2$$

$$\log_e 3\text{·}241 \quad - \quad - \quad 1\text{·}1759$$
$$\log_e 10^2 \quad - \quad - \quad 4\text{·}6052$$

$$\log_e 324\text{·}1 \quad - \quad - \quad 5\text{·}7811$$

$$(3) \quad \log_e \text{·}0003241 = \log_e 3\text{·}241 \times 10^{-4}$$

$$\log_e 3\text{·}241 \quad - \quad - \quad 1\text{·}1759$$
$$\log_e 10^{-4} \quad - \quad - \quad \overline{10}\text{·}7897$$

$$\log_e \text{·}0003241 \quad - \quad - \quad \overline{9}\text{·}9656$$

In calculating the value of a fraction, of which the numerator and denominator each consists of two or more factors, it is often of advantage, instead of *subtracting* the logarithms of the denominator factors, to add in the logarithms of their reciprocals—the complemental logarithms or co-logs as they are sometimes called.

* That which causes the necessity for two such tables gives rise also to the necessity for watchfulness on the part of the calculator. Probably the best preventive against mistakes is the habit, excellent in all calculations, of making a mental estimate of the number to be expected as the result of taking the square root. Mistakes may also be avoided easily and with little loss of time by comparing with the table of squares.

RULE X. To find the mantissa of the logarithm of a reciprocal.—Write down the difference between the mantissa of the logarithm of the given number and 1·0000 ; or simply, commencing at the left hand, write down the series of numbers which will make each figure of the mantissa of the logarithm of the number up to be equal to 9, except the last significant figure, which must be made up to 10.

RULE XI. Otherwise : To find the mantissa of the logarithm of a reciprocal.—Proceed as in Rule III., using the Table of Logarithms of Reciprocals.

RULE XII. To find the index for the logarithm of a reciprocal.—If the given number consist of a whole number and a decimal, the index is equal to the number of the digits which constitute the whole number, and is *negative*. If the given number is a decimal without a whole number the index is equal to the number of zeros which follow the decimal point and is *positive*.

Examples. Find

$$\log \frac{1}{237\cdot4}$$

Mantissa of co-log from table, p. 16.   -   -   ·6246.

Index   -   -   -   -   -   -   - -3

$$\log \frac{1}{237\cdot4} = \cdot6246 - 3 \text{ or } \bar{3}\cdot6246$$

$$\log \frac{1}{\cdot002374} = \cdot6246 + 2 \text{ or } 2\cdot6246$$

Remark.—In finding the logarithm of the reciprocal of a trigonometrical function it is only necessary to subtract the *tabular logarithm* from 10.   This will readily be seen from an example.

Since (see *Note* to Rule VII.),

$$\log \sin 36° = \text{Tab. log } \sin 36° - 10$$

we have

$$\log \left(\frac{1}{\sin 36°} \right) = -\log \sin 36° = 10 - \text{Tab. log } \sin 36°.$$

The subtraction from 10 is most easily performed by writing down the numbers which make up the figures of the tabular logarithm to 9, as in Rule X., except in the case of the last significant figure, for which write the number which, if added to it, would make it up to be 10.

An example of calculation is given here in order to show a convenient way of writing down the given numbers and their logarithms. It is scarcely possible to overestimate the importance of strict adherence to method ; for instance, in physical calculations.   In the first place errors are thus most easily avoided or detected ; and it is also frequently useful to be able to return on the arithmetical steps in order to make an alteration of form, or, if improved data are forthcoming, to obtain a result true to a closer degree of approximation.

Example. Calculate the value of

$$\frac{27\cdot34 \times 0\cdot1325 \times \sin 29°}{14\cdot23 \times \cdot00176 \times \tan 34°}$$

Numbers.						Logs.
27·34	-	-	-	-	-	·4368 + 1
0·1325 (see table, p. 12)				-		·1222 − 1
sin 29°	-	-	-	-	-	9·6856 − 10
1/14·23	-	-	-		-	·8468 − 2
1/·00176	-	-	-	-	-	·7545 + 2
1/tan 34°	-	-	-	-	-	0·1710 + 0
						·0169 + 2
Result	-	-		1·040 × 10².		

In bringing out a Second Edition of this Book of Tables, I desire to acknowledge the kind assistance of friends ; and in particular the valuable criticisms and suggestions which I have received from Prof. Sir G. Gabriel Stokes, Bart., Pres. R. S., and from Prof. G. H. Darwin, F.R.S.

J. T. B.

February 18, 1890.

	0	1	2	3	4	5	6	7	8	9
100	0000	0004	0009	0013	0017	0022	0026	0030	0035	0039
101	0043	0048	0052	0056	0060	0065	0069	0073	0077	0082
102	0086	0090	0095	0099	0103	0107	0111	0116	0120	0124
103	0128	0133	0137	0141	0145	0149	0154	0158	0162	0166
104	0170	0175	0179	0183	0187	0191	0195	0199	0204	0208
105	0212	0216	0220	0224	0228	0233	0237	0241	0245	0249
106	0253	0257	0261	0265	0269	0273	0278	0282	0286	0290
107	0294	0298	0302	0306	0310	0314	0318	0322	0326	0330
108	0334	0338	0342	0346	0350	0354	0358	0362	0366	0370
109	0374	0378	0382	0386	0390	0394	0398	0402	0406	0410
110	0414	0418	0422	0426	0430	0434	0438	0441	0445	0449
111	0453	0457	0461	0465	0469	0473	0477	0481	0484	0488
112	0492	0496	0500	0504	0508	0512	0515	0519	0523	0527
113	0531	0535	0538	0542	0546	0550	0554	0558	0561	0565
114	0569	0573	0577	0580	0584	0588	0592	0596	0599	0603
115	0607	0611	0615	0618	0622	0626	0630	0633	0637	0641
116	0645	0648	0652	0656	0660	0663	0667	0671	0674	0678
117	0682	0686	0689	0693	0697	0700	0704	0708	0711	0715
118	0719	0722	0726	0730	0734	0737	0741	0745	0748	0752
119	0755	0759	0763	0766	0770	0774	0777	0781	0785	0788
120	0792	0795	0799	0803	0806	0810	0813	0817	0821	0824
121	0828	0831	0835	0839	0842	0846	0849	0853	0856	0860
122	0864	0867	0871	0874	0878	0881	0885	0888	0892	0896
123	0899	0903	0906	0910	0913	0917	0920	0924	0927	0931
124	0934	0938	0941	0945	0948	0952	0955	0959	0962	0966
125	0969	0973	0976	0980	0983	0986	0990	0993	0997	1000
126	1004	1007	1011	1014	1017	1021	1024	1028	1031	1035
127	1038	1041	1045	1048	1052	1055	1059	1062	1065	1069
128	1072	1075	1079	1082	1086	1089	1092	1096	1099	1103
129	1106	1109	1113	1116	1119	1123	1126	1129	1133	1136
130	1139	1143	1146	1149	1153	1156	1159	1163	1166	1169
131	1173	1176	1179	1183	1186	1189	1193	1196	1199	1202
132	1206	1209	1212	1216	1219	1222	1225	1229	1232	1235
133	1239	1242	1245	1248	1252	1255	1258	1261	1265	1268
134	1271	1274	1278	1281	1284	1287	1290	1294	1297	1300
135	1303	1307	1310	1313	1316	1319	1323	1326	1329	1332
136	1335	1339	1342	1345	1348	1351	1355	1358	1361	1364
137	1367	1370	1374	1377	1380	1383	1386	1389	1392	1396
138	1399	1402	1405	1408	1411	1414	1418	1421	1424	1427
139	1430	1433	1436	1440	1443	1446	1449	1452	1455	1458
140	1461	1464	1467	1471	1474	1477	1480	1483	1486	1489

	0	1	2	3	4	5	6	7	8	9
100		9996	9991	9987	9983	9978	9974	9970	9965	9961
101	9957	9952	9948	9944	9940	9935	9931	9927	9923	9918
102	9914	9910	9905	9901	9897	9893	9889	9884	9880	9876
103	9872	9867	9863	9859	9855	9851	9846	9842	9838	9834
104	9830	9825	9821	9817	9813	9809	9805	9801	9796	9792
105	9788	9784	9780	9776	9772	9767	9763	9759	9755	9751
106	9747	9743	9739	9735	9731	9727	9722	9718	9714	9710
107	9706	9702	9698	9694	9690	9686	9682	9678	9674	9670
108	9666	9662	9658	9654	9650	9646	9642	9638	9634	9630
109	9626	9622	9618	9614	9610	9606	9602	9598	9594	9590
110	9586	9582	9578	9574	9570	9566	9562	9559	9555	9551
111	9547	9543	9539	9535	9531	9527	9523	9519	9516	9512
112	9508	9504	9500	9496	9492	9488	9485	9481	9477	9473
113	9469	9465	9462	9458	9454	9450	9446	9442	9439	9435
114	9431	9427	9423	9420	9416	9412	9408	9404	9401	9397
115	9393	9389	9385	9382	9378	9374	9370	9367	9363	9359
116	9355	9352	9348	9344	9340	9337	9333	9329	9326	9322
117	9318	9314	9311	9307	9303	9300	9296	9292	9289	9285
118	9281	9278	9274	9270	9266	9263	9259	9255	9252	9248
119	9245	9241	9237	9234	9230	9226	9223	9219	9215	9212
120	9208	9205	9201	9197	9194	9190	9187	9183	9179	9176
121	9172	9169	9165	9161	9158	9154	9151	9147	9144	9140
122	9136	9133	9129	9126	9122	9119	9115	9112	9108	9104
123	9101	9097	9094	9090	9087	9083	9080	9076	9073	9069
124	9066	9062	9059	9055	9052	9048	9045	9041	9038	9034
125	9031	9027	9024	9020	9017	9014	9010	9007	9003	9000
126	8996	8993	8989	8986	8983	8979	8976	8972	8969	8965
127	8962	8959	8955	8952	8948	8945	8941	8938	8935	8931
128	8928	8925	8921	8918	8914	8911	8908	8904	8901	8897
129	8894	8891	8887	8884	8881	8877	8874	8871	8867	8864
130	8861	8857	8854	8851	8847	8844	8841	8837	8834	8831
131	8827	8824	8821	8817	8814	8811	8807	8804	8801	8798
132	8794	8791	8788	8784	8781	8778	8775	8771	8768	8765
133	8761	8758	8755	8752	8748	8745	8742	8739	8735	8732
134	8729	8726	8722	8719	8716	8713	8710	8706	8703	8700
135	8697	8693	8690	8687	8684	8681	8677	8674	8671	8668
136	8665	8661	8658	8655	8652	8649	8645	8642	8639	8636
137	8633	8630	8626	8623	8620	8617	8614	8611	8608	8604
138	8601	8598	8595	8592	8589	8586	8582	8579	8576	8573
139	8570	8567	8564	8560	8557	8554	8551	8548	8545	8542
140	8539	8536	8533	8529	8526	8523	8520	8517	8514	8511

	0	1	2	3	4	5	6	7	8	9	1	2	3	4	5	6	7	8	9
10	.0000	0043	0086	0128	0170	0212	0253	0294	0334	0374	4	8	12	17	21	25	29	33	37
11	0414	0453	0492	0531	0569	0607	0645	0682	0719	0755	4	8	11	15	19	23	26	30	34
12	0792	0828	0864	0899	0934	0969	1004	1038	1072	1106	3	7	10	14	17	21	24	28	31
13	1139	1173	1206	1239	1271	1303	1335	1367	1399	1430	3	6	10	13	16	19	23	26	29
14	1461	1492	1523	1553	1584	1614	1644	1673	1703	1732	3	6	9	12	15	18	21	24	27
15	1761	1790	1818	1847	1875	1903	1931	1959	1987	2014	3	6	8	11	14	17	20	22	25
16	2041	2068	2095	2122	2148	2175	2201	2227	2253	2279	3	5	8	11	13	16	18	21	24
17	2304	2330	2355	2380	2405	2430	2455	2480	2504	2529	2	5	7	10	12	15	17	20	22
18	2553	2577	2601	2625	2648	2672	2695	2718	2742	2765	2	5	7	9	12	14	16	19	21
19	2788	2810	2833	2856	2878	2900	2923	2945	2967	2989	2	4	7	9	11	13	16	18	20
20	3010	3032	3054	3075	3096	3118	3139	3160	3181	3201	2	4	6	8	11	13	15	17	19
21	3222	3243	3263	3284	3304	3324	3345	3365	3385	3404	2	4	6	8	10	12	14	16	18
22	3424	3444	3464	3483	3502	3522	3541	3560	3579	3598	2	4	6	8	10	12	14	15	17
23	3617	3636	3655	3674	3692	3711	3729	3747	3766	3784	2	4	6	7	9	11	13	15	17
24	3802	3820	3838	3856	3874	3892	3909	3927	3945	3962	2	4	5	7	9	11	12	14	16
25	3979	3997	4014	4031	4048	4065	4082	4099	4116	4133	2	3	5	7	9	10	12	14	15
26	4150	4166	4183	4200	4216	4232	4249	4265	4281	4298	2	3	5	7	8	10	11	13	15
27	4314	4330	4346	4362	4378	4393	4409	4425	4440	4456	2	3	5	6	8	9	11	13	14
28	4472	4487	4502	4518	4533	4548	4564	4579	4594	4609	2	3	5	6	8	9	11	12	14
29	4624	4639	4654	4669	4683	4698	4713	4728	4742	4757	1	3	4	6	7	9	10	12	13
30	4771	4786	4800	4814	4829	4843	4857	4871	4886	4900	1	3	4	6	7	9	10	11	13
31	4914	4928	4942	4955	4969	4983	4997	5011	5024	5038	1	3	4	6	7	8	10	11	12
32	5051	5065	5079	5092	5105	5119	5132	5145	5159	5172	1	3	4	5	7	8	9	11	12
33	5185	5198	5211	5224	5237	5250	5263	5276	5289	5302	1	3	4	5	6	8	9	10	12
34	5315	5328	5340	5353	5366	5378	5391	5403	5416	5428	1	3	4	5	6	8	9	10	11
35	5441	5453	5465	5478	5490	5502	5514	5527	5539	5551	1	2	4	5	6	7	9	10	11
36	5563	5575	5587	5599	5611	5623	5635	5647	5658	5670	1	2	4	5	6	7	8	10	11
37	5682	5694	5705	5717	5729	5740	5752	5763	5775	5786	1	2	3	5	6	7	8	9	10
38	5798	5809	5821	5832	5843	5855	5866	5877	5888	5899	1	2	3	5	6	7	8	9	10
39	5911	5922	5933	5944	5955	5966	5977	5988	5999	6010	1	2	3	4	5	7	8	9	10
40	6021	6031	6042	6053	6064	6075	6085	6096	6107	6117	1	2	3	4	5	6	8	9	10
41	6128	6138	6149	6160	6170	6180	6191	6201	6212	6222	1	2	3	4	5	6	7	8	9
42	6232	6243	6253	6263	6274	6284	6294	6304	6314	6325	1	2	3	4	5	6	7	8	9
43	6335	6345	6355	6365	6375	6385	6395	6405	6415	6425	1	2	3	4	5	6	7	8	9
44	6435	6444	6454	6464	6474	6484	6493	6503	6513	6522	1	2	3	4	5	6	7	8	9
45	6532	6542	6551	6561	6571	6580	6590	6599	6609	6618	1	2	3	4	5	6	7	8	9
46	6628	6637	6646	6656	6665	6675	6684	6693	6702	6712	1	2	3	4	5	6	7	7	8
47	6721	6730	6739	6749	6758	6767	6776	6785	6794	6803	1	2	3	4	5	5	6	7	8
48	6812	6821	6830	6839	6848	6857	6866	6875	6884	6893	1	2	3	4	4	5	6	7	8
49	6902	6911	6920	6928	6937	6946	6955	6964	6972	6981	1	2	3	4	4	5	6	7	8
50	6990	6998	7007	7016	7024	7033	7042	7050	7059	7067	1	2	3	3	4	5	6	7	8
51	7076	7084	7093	7101	7110	7118	7126	7135	7143	7152	1	2	3	3	4	5	6	7	8
52	7160	7168	7177	7185	7193	7202	7210	7218	7226	7235	1	2	2	3	4	5	6	7	7
53	7243	7251	7259	7267	7275	7284	7292	7300	7308	7316	1	2	2	3	4	5	6	6	7
54	7324	7332	7340	7348	7356	7364	7372	7380	7388	7396	1	2	2	3	4	5	6	6	7

	4	5	6	7	8	9	1	2	3	4	5	6	7	8	9
7	7435	7443	7451	7459	7466	7474	1	2	2	3	4	5	5	6	7
5	7513	7520	7528	7536	7543	7551	1	2	2	3	4	5	5	6	
2	7589	7597	7604	7612	7619	7627	1	2	2	3	4	5	5	6	7
7	7664	7672	7679	7686	7694	7701	1	1	2	3	4	4	5	6	7
1	7738	7745	7752	7760	7767	7774	1	1	2	3	4	4	5	6	7
3	7810	7818	7825	7832	7839	7846	1	1	2	3	4	4	5	6	6
5	7882	7889	7896	7903	7910	7917	1	1	2	3	4	4	5	6	6
5	7952	7959	7966	7973	7980	7987	1	1	2	3	3	4	5	6	6
4	8021	8028	8035	8041	8048	8055	1	1	2	3	3	4	5	5	6
2	8089	8096	8102	8109	8116	8122	1	1	2	3	3	4	5	5	6
9	8156	8162	8169	8176	8182	8189	1	1	2	3	3	4	5	5	6
5	8222	8228	8235	8241	8248	8254	1	1	2	3	3	4	5	5	6
0	8287	8293	8299	8306	8312	8319	1	1	2	3	3	4	5	5	6
4	8351	8357	8363	8370	8376	8382	1	1	2	3	3	4	4	5	6
7	8414	8420	8426	8432	8439	8445	1	1	2	2	3	4	4	5	6
0	8476	8482	8488	8494	8500	8506	1	1	2	2	3	4	4	5	6
1	8537	8543	8549	8555	8561	8567	1	1	2	2	3	4	4	5	5
1	8597	8603	8609	8615	8621	8627	1	1	2	2	3	4	4	5	5
1	8657	8663	8669	8675	8681	8686	1	1	2	2	3	4	4	5	5
9	8716	8722	8727	8733	8739	8745	1	1	2	2	3	4	4	5	5
8	8774	8779	8785	8791	8797	8802	1	1	2	2	3	3	4	5	5
5	8831	8837	8842	8848	8854	8859	1	1	2	2	3	3	4	5	5
2	8887	8893	8899	8904	8910	8915	1	1	2	2	3	3	4	4	5
8	8943	8949	8954	8960	8965	8971	1	1	2	2	3	3	4	4	5
3	8998	9004	9009	9015	9020	9025	1	1	2	2	3	3	4	4	5
7	9053	9058	9063	9069	9074	9079	1	1	2	2	3	3	4	4	5
1	9106	9112	9117	9122	9128	9133	1	1	2	2	3	3	4	4	5
4	9159	9165	9170	9175	9180	9186	1	1	2	2	3	3	4	4	5
6	9212	9217	9222	9227	9232	9238	1	1	2	2	3	3	4	4	5
8	9263	9269	9274	9279	9284	9289	1	1	2	2	3	3	4	4	5
9	9315	9320	9325	9330	9335	9340	1	1	2	2	3	3	4	4	5
0	9365	9370	9375	9380	9385	9390	1	1	2	2	3	3	4	4	5
0	9415	9420	9425	9430	9435	9440	0	1	1	2	2	3	3	4	4
0	9465	9469	9474	9479	9484	9489	0	1	1	2	2	3	3	4	4
9	9513	9518	9523	9528	9533	9538	0	1	1	2	2	3	3	4	4
7	9562	9566	9571	9576	9581	9586	0	1	1	2	2	3	3	4	4
5	9609	9614	9619	9624	9628	9633	0	1	1	2	2	3	3	4	4
2	9657	9661	9666	9671	9675	9680	0	1	1	2	2	3	3	4	4
9	9703	9708	9713	9717	9722	9727	0	1	1	2	2	3	3	4	4
5	9750	9754	9759	9763	9768	9773	0	1	1	2	2	3	3	4	4
1	9795	9800	9805	9809	9814	9818	0	1	1	2	2	3	3	4	4
6	9841	9845	9850	9854	9859	9863	0	1	1	2	2	3	3	4	4
1	9886	9890	9894	9899	9903	9908	0	1	1	2	2	3	3	4	4
6	9930	9934	9939	9943	9948	9952	0	1	1	2	2	3	3	4	4
9	9974	9978	9983	9987	9991	9996	0	1	1	2	2	3	3	3	4

LOGARITHMS OF RECIPROCALS.

	0	1	2	3	4	5	6	7	8	9	1	2	3	4	5	6	7	8	9
10		9957	9914	9872	9830	9788	9747	9706	9666	9626	4	8	12	17	21	25	29	33	37
11	9586	9547	9508	9469	9431	9393	9355	9318	9281	9245	4	8	11	15	19	23	26	30	34
12	9208	9172	9136	9101	9066	9031	8996	8962	8928	8894	3	7	10	14	17	21	24	28	31
13	8861	8827	8794	8761	8729	8697	8665	8633	8601	8570	3	6	10	13	16	19	23	26	29
14	8539	8508	8477	8447	8416	8386	8356	8327	8297	8268	3	6	9	12	15	18	21	24	27
15	8239	8210	8182	8153	8125	8097	8069	8041	8013	7986	3	6	8	11	14	17	20	22	25
16	7959	7932	7905	7878	7852	7825	7799	7773	7747	7721	3	5	8	11	13	16	18	21	24
17	7696	7670	7645	7620	7595	7570	7545	7520	7496	7471	2	5	7	10	12	15	17	20	22
18	7447	7423	7399	7375	7352	7328	7305	7282	7258	7235	2	5	7	9	12	14	16	19	21
19	7212	7190	7167	7144	7122	7100	7077	7055	7033	7011	2	4	7	9	11	13	16	18	20
20	6990	6968	6946	6925	6904	6882	6861	6840	6819	6799	2	4	6	8	11	13	15	17	19
21	6778	6757	6737	6716	6696	6676	6655	6635	6615	6596	2	4	6	8	10	12	14	16	18
22	6576	6556	6536	6517	6497	6478	6459	6440	6421	6402	2	4	6	8	10	12	14	15	17
23	6383	6364	6345	6326	6308	6289	6271	6253	6234	6216	2	4	6	7	9	11	13	15	17
24	6198	6180	6162	6144	6126	6108	6091	6073	6055	6038	2	4	5	7	9	11	12	14	16
25	6021	6003	5986	5969	5952	5935	5918	5901	5884	5867	2	3	5	7	9	10	12	14	15
26	5850	5834	5817	5800	5784	5768	5751	5735	5719	5702	2	3	5	7	8	10	11	13	15
27	5686	5670	5654	5638	5622	5607	5591	5575	5560	5544	2	3	5	6	8	9	11	13	14
28	5528	5513	5498	5482	5467	5452	5436	5421	5406	5391	2	3	5	6	8	9	11	12	14
29	5376	5361	5346	5331	5317	5302	5287	5272	5258	5243	1	3	4	6	7	9	10	12	13
30	5229	5214	5200	5186	5171	5157	5143	5129	5114	5100	1	3	4	6	7	9	10	11	13
31	5086	5072	5058	5045	5031	5017	5003	4989	4976	4962	1	3	4	6	7	8	10	11	12
32	4948	4935	4921	4908	4895	4881	4868	4855	4841	4828	1	3	4	5	7	8	9	11	12
33	4815	4802	4789	4776	4763	4750	4737	4724	4711	4698	1	3	4	5	6	8	9	10	12
34	4685	4672	4660	4647	4634	4622	4609	4597	4584	4572	1	2	4	5	6	8	9	10	11
35	4559	4547	4535	4522	4510	4498	4486	4473	4461	4449	1	2	4	5	6	7	9	10	11
36	4437	4425	4413	4401	4389	4377	4365	4353	4342	4330	1	2	4	5	6	7	9	10	11
37	4318	4306	4295	4283	4271	4260	4248	4237	4225	4214	1	2	3	5	6	7	8	9	10
38	4202	4191	4179	4168	4157	4145	4134	4123	4112	4101	1	2	3	5	6	7	8	9	10
39	4089	4078	4067	4056	4045	4034	4023	4012	4001	3990	1	2	3	4	6	7	8	9	10
40	3979	3969	3958	3947	3936	3925	3915	3904	3893	3883	1	2	3	4	5	6	7	8	9
41	3872	3862	3851	3840	3830	3820	3809	3799	3788	3778	1	2	3	4	5	6	7	8	9
42	3768	3757	3747	3737	3726	3716	3706	3696	3686	3675	1	2	3	4	5	6	7	8	9
43	3665	3655	3645	3635	3625	3615	3605	3595	3585	3575	1	2	3	4	5	6	7	8	9
44	3565	3556	3546	3536	3526	3516	3507	3497	3487	3478	1	2	3	4	5	6	7	8	9
45	3468	3458	3449	3439	3429	3420	3410	3401	3391	3382	1	2	3	4	5	6	7	8	9
46	3372	3363	3354	3344	3335	3325	3316	3307	3298	3288	1	2	3	4	5	6	7	7	8
47	3279	3270	3261	3251	3242	3233	3224	3215	3206	3197	1	2	3	4	5	5	6	7	8
48	3188	3179	3170	3161	3152	3143	3134	3125	3116	3107	1	2	3	4	4	5	6	7	8
49	3098	3089	3080	3071	3063	3054	3045	3036	3028	3019	1	2	3	4	4	5	6	7	8
50	3010	3002	2993	2984	2976	2967	2958	2950	2941	2933	1	2	3	3	4	5	6	7	8
51	2924	2916	2907	2899	2890	2882	2874	2865	2857	2848	1	2	3	3	4	5	6	7	8
52	2840	2832	2823	2815	2807	2798	2790	2782	2774	2765	1	2	2	3	4	5	6	7	7
53	2757	2749	2741	2733	2725	2716	2708	2700	2692	2684	1	2	2	3	4	5	6	6	7
54	2676	2668	2660	2652	2644	2636	2628	2620	2612	2604	1	2	2	3	4	5	6	6	7

N.B.—Numbers in Difference Columns to be Subtracted, not Added.

	0	1	2	3	4	5	6	7	8	9	1	2	3	4	5	6	7	8	9
55	2596	2588	2581	2573	2565	2557	2549	2541	2534	2526	1	2	2	3	4	5	5	6	7
56	2518	2510	2503	2495	2487	2480	2472	2464	2457	2449	1	2	2	3	4	5	5	6	7
57	2441	2434	2426	2418	2411	2403	2396	2388	2381	2373	1	2	2	3	4	5	5	6	7
58	2366	2358	2351	2343	2336	2328	2321	2314	2306	2299	1	1	2	3	4	4	5	6	7
59	2291	2284	2277	2269	2262	2255	2248	2240	2233	2226	1	1	2	3	4	4	5	6	7
60	2218	2211	2204	2197	2190	2182	2175	2168	2161	2154	1	1	2	3	4	4	5	6	6
61	2147	2140	2132	2125	2118	2111	2104	2097	2090	2083	1	1	2	3	4	4	5	6	6
62	2076	2069	2062	2055	2048	2041	2034	2027	2020	2013	1	1	2	3	3	4	5	6	6
63	2007	2000	1993	1986	1979	1972	1965	1959	1952	1945	1	1	2	3	3	4	5	5	6
64	1938	1931	1925	1918	1911	1904	1898	1891	1884	1878	1	1	2	3	3	4	5	5	6
65	1871	1864	1858	1851	1844	1838	1831	1824	1818	1811	1	1	2	3	3	4	5	5	6
66	1805	1798	1791	1785	1778	1772	1765	1759	1752	1746	1	1	2	3	3	4	5	5	6
67	1739	1733	1726	1720	1713	1707	1701	1694	1688	1681	1	1	2	3	3	4	4	5	6
68	1675	1669	1662	1656	1649	1643	1637	1630	1624	1618	1	1	2	3	3	4	4	5	6
69	1612	1605	1599	1593	1586	1580	1574	1568	1561	1555	1	1	2	3	3	4	4	5	6
70	1549	1543	1537	1530	1524	1518	1512	1506	1500	1494	1	1	2	2	3	4	4	5	5
71	1487	1481	1475	1469	1463	1457	1451	1445	1439	1433	1	1	2	2	3	4	4	5	5
72	1427	1421	1415	1409	1403	1397	1391	1385	1379	1373	1	1	2	2	3	4	4	5	5
73	1367	1361	1355	1349	1343	1337	1331	1325	1319	1314	1	1	2	2	3	4	4	5	5
74	1308	1302	1296	1290	1284	1278	1273	1267	1261	1255	1	1	2	2	3	3	4	5	5
75	1249	1244	1238	1232	1226	1221	1215	1209	1203	1198	1	1	2	2	3	3	4	5	5
76	1192	1186	1180	1175	1169	1163	1158	1152	1146	1141	1	1	2	2	3	3	4	5	5
77	1135	1129	1124	1118	1113	1107	1101	1096	1090	1085	1	1	2	2	3	3	4	4	5
78	1079	1073	1068	1062	1057	1051	1046	1040	1035	1029	1	1	2	2	3	3	4	4	5
79	1024	1018	1013	1007	1002	0996	0991	0985	0980	0975	1	1	2	2	3	3	4	4	5
80	0969	0964	0958	0953	0947	0942	0937	0931	0926	0921	1	1	2	2	3	3	4	4	5
81	0915	0910	0904	0899	0894	0888	0883	0878	0872	0867	1	1	2	2	3	3	4	4	5
82	0862	0857	0851	0846	0841	0835	0830	0825	0820	0814	1	1	2	2	3	3	4	4	5
83	0809	0804	0799	0794	0788	0783	0778	0773	0768	0762	1	1	2	2	3	3	4	4	5
84	0757	0752	0747	0742	0737	0731	0726	0721	0716	0711	1	1	2	2	3	3	4	4	5
85	0706	0701	0696	0691	0685	0680	0675	0670	0665	0660	1	1	2	2	3	3	4	4	5
86	0655	0650	0645	0640	0635	0630	0625	0620	0615	0610	1	1	2	2	3	3	4	4	5
87	0605	0600	0595	0590	0585	0580	0575	0570	0565	0560	0	1	1	2	2	3	3	4	4
88	0555	0550	0545	0540	0535	0531	0526	0521	0516	0511	0	1	1	2	2	3	3	4	4
89	0506	0501	0496	0491	0487	0482	0477	0472	0467	0462	0	1	1	2	2	3	3	4	4
90	0458	0453	0448	0443	0438	0434	0429	0424	0419	0414	0	1	1	2	2	3	3	4	4
91	0410	0405	0400	0395	0391	0386	0381	0376	0372	0367	0	1	1	2	2	3	3	4	4
92	0362	0357	0353	0348	0343	0339	0334	0329	0325	0320	0	1	1	2	2	3	3	4	4
93	0315	0311	0306	0301	0297	0292	0287	0283	0278	0273	0	1	1	2	2	3	3	4	4
94	0269	0264	0259	0255	0250	0246	0241	0237	0232	0227	0	1	1	2	2	3	3	4	4
95	0223	0218	0214	0209	0205	0200	0195	0191	0186	0182	0	1	1	2	2	3	3	4	4
96	0177	0173	0168	0164	0159	0155	0150	0146	0141	0137	0	1	1	2	2	3	3	4	4
97	0132	0128	0123	0119	0114	0110	0106	0101	0097	0092	0	1	1	2	2	3	3	4	4
98	0088	0083	0079	0074	0070	0066	0061	0057	0052	0048	0	1	1	2	2	3	3	4	4
99	0044	0039	0035	0031	0026	0022	0017	0013	0009	0004	0	1	1	2	2	3	3	3	4

N.B.—Numbers in Difference Columns to be Subtracted, not Added.

	0	1	2	3	4	5	6	7	8	9	1	2	3	4	5	6	7	8	9
·00	1000	1002	1005	1007	1009	1012	1014	1016	1019	1021	0	0	1	1	1	1	2	2	2
·01	1023	1026	1028	1030	1033	1035	1038	1040	1042	1045	0	0	1	1	1	1	2	2	2
·02	1047	1050	1052	1054	1057	1059	1062	1064	1067	1069	0	0	1	1	1	1	2	2	2
·03	1072	1074	1076	1079	1081	1084	1086	1089	1091	1094	0	0	1	1	1	1	2	2	2
·04	1096	1099	1102	1104	1107	1109	1112	1114	1117	1119	0	1	1	1	1	2	2	2	2
·05	1122	1125	1127	1130	1132	1135	1138	1140	1143	1146	0	1	1	1	1	2	2	2	2
·06	1148	1151	1153	1156	1159	1161	1164	1167	1169	1172	0	1	1	1	1	2	2	2	2
·07	1175	1178	1180	1183	1186	1189	1191	1194	1197	1199	0	1	1	1	1	2	2	2	2
·08	1202	1205	1208	1211	1213	1216	1219	1222	1225	1227	0	1	1	1	1	2	2	2	3
·09	1230	1233	1236	1239	1242	1245	1247	1250	1253	1256	0	1	1	1	1	2	2	2	3
·10	1259	1262	1265	1268	1271	1274	1276	1279	1282	1285	0	1	1	1	1	2	2	2	3
·11	1288	1291	1294	1297	1300	1303	1306	1309	1312	1315	0	1	1	1	2	2	2	2	3
·12	1318	1321	1324	1327	1330	1334	1337	1340	1343	1346	0	1	1	1	2	2	2	2	3
·13	1349	1352	1355	1358	1361	1365	1368	1371	1374	1377	0	1	1	1	2	2	2	3	3
·14	1380	1384	1387	1390	1393	1396	1400	1403	1406	1409	0	1	1	1	2	2	2	3	3
·15	1413	1416	1419	1422	1426	1429	1432	1435	1439	1442	0	1	1	1	2	2	2	3	3
·16	1445	1449	1452	1455	1459	1462	1466	1469	1472	1476	0	1	1	1	2	2	2	3	3
·17	1479	1483	1486	1489	1493	1496	1500	1503	1507	1510	0	1	1	1	2	2	2	3	3
·18	1514	1517	1521	1524	1528	1531	1535	1538	1542	1545	0	1	1	1	2	2	2	3	3
·19	1549	1552	1556	1560	1563	1567	1570	1574	1578	1581	0	1	1	1	2	2	3	3	3
·20	1585	1589	1592	1596	1600	1603	1607	1611	1614	1618	0	1	1	1	2	2	3	3	3
·21	1622	1626	1629	1633	1637	1641	1644	1648	1652	1656	0	1	1	2	2	2	3	3	3
·22	1660	1663	1667	1671	1675	1679	1683	1687	1690	1694	0	1	1	2	2	2	3	3	3
·23	1698	1702	1706	1710	1714	1718	1722	1726	1730	1734	0	1	1	2	2	2	3	3	4
·24	1738	1742	1746	1750	1754	1758	1762	1766	1770	1774	0	1	1	2	2	2	3	3	4
·25	1778	1782	1786	1791	1795	1799	1803	1807	1811	1816	0	1	1	2	2	2	3	3	4
·26	1820	1824	1828	1832	1837	1841	1845	1849	1854	1858	0	1	1	2	2	3	3	3	4
·27	1862	1866	1871	1875	1879	1884	1888	1892	1897	1901	0	1	1	2	2	3	3	3	4
·28	1905	1910	1914	1919	1923	1928	1932	1936	1941	1945	0	1	1	2	2	3	3	4	4
·29	1950	1954	1959	1963	1968	1972	1977	1982	1986	1991	0	1	1	2	2	3	3	4	4
·30	1995	2000	2004	2009	2014	2018	2023	2028	2032	2037	0	1	1	2	2	3	3	4	4
·31	2042	2046	2051	2056	2061	2065	2070	2075	2080	2084	0	1	1	2	2	3	3	4	4
·32	2089	2094	2099	2104	2109	2113	2118	2123	2128	2133	0	1	1	2	2	3	3	4	4
·33	2138	2143	2148	2153	2158	2163	2168	2173	2178	2183	0	1	1	2	2	3	3	4	4
·34	2188	2193	2198	2203	2208	2213	2218	2223	2228	2234	1	1	2	2	3	3	4	4	5
·35	2239	2244	2249	2254	2259	2265	2270	2275	2280	2286	1	1	2	2	3	3	4	4	5
·36	2291	2296	2301	2307	2312	2317	2323	2328	2333	2339	1	1	2	2	3	3	4	4	5
·37	2344	2350	2355	2360	2366	2371	2377	2382	2388	2393	1	1	2	2	3	3	4	4	5
·38	2399	2404	2410	2415	2421	2427	2432	2438	2443	2449	1	1	2	2	3	3	4	4	5
·39	2455	2460	2466	2472	2477	2483	2489	2495	2500	2506	1	1	2	2	3	3	4	5	5
·40	2512	2518	2523	2529	2535	2541	2547	2553	2559	2564	1	1	2	2	3	4	4	5	5
·41	2570	2576	2582	2588	2594	2600	2606	2612	2618	2624	1	1	2	2	3	4	4	5	5
·42	2630	2636	2642	2649	2655	2661	2667	2673	2679	2685	1	1	2	2	3	4	4	5	6
·43	2692	2698	2704	2710	2716	2723	2729	2735	2742	2748	1	1	2	3	3	4	4	5	6
·44	2754	2761	2767	2773	2780	2786	2793	2799	2805	2812	1	1	2	3	3	4	4	5	6
·45	2818	2825	2831	2838	2844	2851	2858	2864	2871	2877	1	1	2	3	3	4	5	5	6
·46	2884	2891	2897	2904	2911	2917	2924	2931	2938	2944	1	1	2	3	3	4	5	5	6
·47	2951	2958	2965	2972	2979	2985	2992	2999	3006	3013	1	1	2	3	3	4	5	5	6
·48	3020	3027	3034	3041	3048	3055	3062	3069	3076	3083	1	1	2	3	4	4	5	6	6
·49	3090	3097	3105	3112	3119	3126	3133	3141	3148	3155	1	1	2	3	4	4	5	6	6

	0	1	2	3	4	5	6	7	8	9	1	2	3	4	5	6	7	8	9
·50	3162	3170	3177	3184	3192	3199	3206	3214	3221	3228	1	1	2	3	4	4	5	6	7
·51	3236	3243	3251	3258	3266	3273	3281	3289	3296	3304	1	2	2	3	4	5	5	6	7
·52	3311	3319	3327	3334	3342	3350	3357	3365	3373	3381	1	2	2	3	4	5	5	6	7
·53	3388	3396	3404	3412	3420	3428	3436	3443	3451	3459	1	2	2	3	4	5	6	6	7
·54	3467	3475	3483	3491	3499	3508	3516	3524	3532	3540	1	2	2	3	4	5	6	6	7
·55	3548	3556	3565	3573	3581	3589	3597	3606	3614	3622	1	2	2	3	4	5	6	7	7
·56	3631	3639	3648	3656	3664	3673	3681	3690	3698	3707	1	2	3	3	4	5	6	7	8
·57	3715	3724	3733	3741	3750	3758	3767	3776	3784	3793	1	2	3	3	4	5	6	7	8
·58	3802	3811	3819	3828	3837	3846	3855	3864	3873	3882	1	2	3	4	4	5	6	7	8
·59	3890	3899	3908	3917	3926	3936	3945	3954	3963	3972	1	2	3	4	5	5	6	7	8
·60	3981	3990	3999	4009	4018	4027	4036	4046	4055	4064	1	2	3	4	5	6	0	7	8
·61	4074	4083	4093	4102	4111	4121	4130	4140	4150	4159	1	2	3	4	5	6	7	8	9
·62	4169	4178	4188	4198	4207	4217	4227	4236	4246	4256	1	2	3	4	5	6	7	8	9
·63	4266	4276	4285	4295	4305	4315	4325	4335	4345	4355	1	2	3	4	5	6	7	8	9
·64	4365	4375	4385	4395	4406	4416	4426	4436	4446	4457	1	2	3	4	5	6	7	8	9
·65	4467	4477	4487	4498	4508	4519	4529	4539	4550	4560	1	2	3	4	5	6	7	8	9
·66	4571	4581	4592	4603	4613	4624	4634	4645	4656	4667	1	2	3	4	5	6	7	9	10
·67	4677	4688	4699	4710	4721	4732	4742	4753	4764	4775	1	2	3	4	5	7	8	9	10
·68	4786	4797	4808	4819	4831	4842	4853	4864	4875	4887	1	2	3	4	6	7	8	9	10
·69	4898	4909	4920	4932	4943	4955	4966	4977	4989	5000	1	2	3	5	6	7	8	9	10
·70	5012	5023	5035	5047	5058	5070	5082	5093	5105	5117	1	2	4	5	6	7	8	9	11
·71	5129	5140	5152	5164	5176	5188	5200	5212	5224	5236	1	2	4	5	6	7	8	10	11
·72	5248	5260	5272	5284	5297	5309	5321	5333	5346	5358	1	2	4	5	6	7	9	10	11
·73	5370	5383	5395	5408	5420	5433	5445	5458	5470	5483	1	3	4	5	6	8	9	10	11
·74	5495	5508	5521	5534	5546	5559	5572	5585	5598	5610	1	3	4	5	6	8	9	10	12
·75	5623	5636	5649	5662	5675	5689	5702	5715	5728	5741	1	3	4	5	7	8	9	10	12
·76	5754	5768	5781	5794	5808	5821	5834	5848	5861	5875	1	3	4	5	7	8	9	11	12
·77	5888	5902	5916	5929	5943	5957	5970	5984	5998	6012	1	3	4	5	7	8	10	11	12
·78	6026	6039	6053	6067	6081	6095	6109	6124	6138	6152	1	3	4	6	7	8	10	11	13
·79	6166	6180	6194	6209	6223	6237	6252	6266	6281	6295	1	3	4	6	7	9	10	11	13
·80	6310	6324	6339	6353	6368	6383	6397	6412	6427	6442	1	3	4	6	7	9	10	12	13
·81	6457	6471	6486	6501	6516	6531	6546	6561	6577	6592	2	3	5	6	8	9	11	12	14
·82	6607	6622	6637	6653	6668	6683	6699	6714	6730	6745	2	3	5	6	8	9	11	12	14
·83	6761	6776	6792	6808	6823	6839	6855	6871	6887	6902	2	3	5	6	8	9	11	13	14
·84	6918	6934	6950	6966	6982	6998	7015	7031	7047	7063	2	3	5	6	8	10	11	13	15
·85	7079	7096	7112	7129	7145	7161	7178	7194	7211	7228	2	3	5	7	8	10	12	13	15
·86	7244	7261	7278	7295	7311	7328	7345	7362	7379	7396	2	3	5	7	8	10	12	13	15
·87	7413	7430	7447	7464	7482	7499	7516	7534	7551	7568	2	3	5	7	9	10	12	14	16
·88	7586	7603	7621	7638	7656	7674	7691	7709	7727	7745	2	4	5	7	9	11	12	14	16
·89	7762	7780	7798	7816	7834	7852	7870	7889	7907	7925	2	4	5	7	9	11	13	14	16
·90	7943	7962	7980	7998	8017	8035	8054	8072	8091	8110	2	4	6	7	9	11	13	15	17
·91	8128	8147	8166	8185	8204	8222	8241	8260	8279	8299	2	4	6	8	9	11	13	15	17
·92	8318	8337	8356	8375	8395	8414	8433	8453	8472	8492	2	4	6	8	10	12	14	15	17
·93	8511	8531	8551	8570	8590	8610	8630	8650	8670	8690	2	4	6	8	10	12	14	16	18
·94	8710	8730	8750	8770	8790	8810	8831	8851	8872	8892	2	4	6	8	10	12	14	16	18
·95	8913	8933	8954	8974	8995	9016	9036	9057	9078	9099	2	4	6	8	10	12	15	17	19
·96	9120	9141	9162	9183	9204	9226	9247	9268	9290	9311	2	4	6	8	11	13	15	17	19
·97	9333	9354	9376	9397	9419	9441	9462	9484	9506	9528	2	4	7	9	11	13	15	17	20
·98	9550	9572	9594	9616	9638	9661	9683	9705	9727	9750	2	4	7	9	11	13	16	18	20
·99	9772	9795	9817	9840	9863	9886	9908	9931	9954	9977	2	5	7	9	11	14	16	18	20

	0'	6'	12'	18'	24'	30'	36'	42'	48'	54'	1	2	3	4	5
0°	Inf. Neg.	7̄·2419	5429	7190	8439	9403	0̄200	0̄870	1̄450	1̄961					
1	8̄·2419	2832	3210	3558	3880	4179	4459	4723	4971	5206					
2	8·5428	5640	5842	6035	6220	6397	6567	6731	6889	7041					
3	8·7188	7330	7468	7602	7731	7857	7979	8098	8213	8326	21	41	62	82	103
4	8·8436	8543	8647	8749	8849	8946	9042	9135	9226	9315	16	32	48	64	80
5	8·9403	9489	9573	9655	9736	9816	9894	9970	0046	0120	13	26	39	52	65
6	9·0192	0264	0334	0403	0472	0539	0605	0670	0734	0797	11	22	33	44	55
7	9·0859	0920	0981	1040	1099	1157	1214	1271	1326	1381	10	19	29	38	48
8	9·1436	1489	1542	1594	1646	1697	1747	1797	1847	1895	8	17	25	34	42
9	9·1943	1991	2038	2085	2131	2176	2221	2266	2310	2353	8	15	23	30	38
10	9·2397	2439	2482	2524	2565	2606	2647	2687	2727	2767	7	14	20	27	34
11	9·2806	2845	2883	2921	2959	2997	3034	3070	3107	3143	6	12	19	25	31
12	9·3179	3214	3250	3284	3319	3353	3387	3421	3455	3488	6	11	17	23	28
13	9·3521	3554	3586	3618	3650	3682	3713	3745	3775	3806	5	11	16	21	26
14	9·3837	3867	3897	3927	3957	3986	4015	4044	4073	4102	5	10	15	20	24
15	9·4130	4158	4186	4214	4242	4269	4296	4323	4350	4377	5	9	14	18	23
16	9·4403	4430	4456	4482	4508	4533	4559	4584	4609	4634	4	9	13	17	21
17	9·4659	4684	4709	4733	4757	4781	4805	4829	4853	4876	4	8	12	16	20
18	9·4900	4923	4946	4969	4992	5015	5037	5060	5082	5104	4	8	11	15	19
19	9·5126	5148	5170	5192	5213	5235	5256	5278	5299	5320	4	7	11	14	18
20	9·5341	5361	5382	5402	5423	5443	5463	5484	5504	5523	3	7	10	14	17
21	9·5543	5563	5583	5602	5621	5641	5660	5679	5698	5717	3	6	10	13	16
22	9·5736	5754	5773	5792	5810	5828	5847	5865	5883	5901	3	6	9	12	15
23	9·5919	5937	5954	5972	5990	6007	6024	6042	6059	6076	3	6	9	12	15
24	9·6093	6110	6127	6144	6161	6177	6194	6210	6227	6243	3	6	8	11	14
25	9·6259	6276	6292	6308	6324	6340	6356	6371	6387	6403	3	5	8	11	13
26	9·6418	6434	6449	6465	6480	6495	6510	6526	6541	6556	3	5	8	10	13
27	9·6570	6585	6600	6615	6629	6644	6659	6673	6687	6702	2	5	7	10	12
28	9·6716	6730	6744	6759	6773	6787	6801	6814	6828	6842	2	5	7	9	12
29	9·6856	6869	6883	6896	6910	6923	6937	6950	6963	6977	2	4	7	9	11
30	9·6990	7003	7016	7029	7042	7055	7068	7080	7093	7106	2	4	6	9	11
31	9·7118	7131	7144	7156	7168	7181	7193	7205	7218	7230	2	4	6	8	10
32	9·7242	7254	7266	7278	7290	7302	7314	7326	7338	7349	2	4	6	8	10
33	9·7361	7373	7384	7396	7407	7419	7430	7442	7453	7464	2	4	6	8	10
34	9·7476	7487	7498	7509	7520	7531	7542	7553	7564	7575	2	4	6	7	9
35	9·7586	7597	7607	7618	7629	7640	7650	7661	7671	7682	2	4	5	7	9
36	9·7692	7703	7713	7723	7734	7744	7754	7764	7774	7785	2	3	5	7	9
37	9·7795	7805	7815	7825	7835	7844	7854	7864	7874	7884	2	3	5	7	8
38	9·7893	7903	7913	7922	7932	7941	7951	7960	7970	7979	2	3	5	6	8
39	9·7989	7998	8007	8017	8026	8035	8044	8053	8063	8072	2	3	5	6	8
40	9·8081	8090	8099	8108	8117	8125	8134	8143	8152	8161	1	3	4	6	7
41	9·8169	8178	8187	8195	8204	8213	8221	8230	8238	8247	1	3	4	6	7
42	9·8255	8264	8272	8280	8289	8297	8305	8313	8322	8330	1	3	4	6	7
43	9·8338	8346	8354	8362	8370	8378	8386	8394	8402	8410	1	3	4	5	7
44	9·8418	8426	8433	8441	8449	8457	8464	8472	8480	8487	1	3	4	5	6

	0′	6′	12′	18′	24′	30′	36′	42′	48′	54′	1	2	3	4	5
45°	9·8495	8502	8510	8517	8525	8532	8540	8547	8555	8562	.1	2	4	5	6
46	9·8569	8577	8584	8591	8598	8606	8613	8620	8627	8634	1	2	4	5	6
47	9·8641	8648	8655	8662	8669	8676	8683	8690	8697	8704	1	2	3	5	6
48	9·8711	8718	8724	8731	8738	8745	8751	8758	8765	8771	1	2	3	4	6
49	9·8778	8784	8791	8797	8804	8810	8817	8823	8830	8836	1	2	3	4	5
50	9·8843	8849	8855	8862	8868	8874	8880	8887	8893	8899	1	2	3	4	5
51	9·8905	8911	8917	8923	8929	8935	8941	8947	8953	8959	1	2	3	4	5
52	9·8965	8971	8977	8983	8989	8995	9000	9006	9012	9018	1	2	3	4	5
53	9·9023	9029	9035	9041	9046	9052	9057	9063	9069	9074	1	2	3	4	5
54	9·9080	9085	9091	9096	9101	9107	9112	9118	9123	9128	1	2	3	4	5
55	9·9134	9139	9144	9149	9155	9160	9165	9170	9175	9181	1	2	3	3	4
56	9·9186	9191	9196	9201	9206	9211	9216	9221	9226	9231	1	2	3	3	4
57	9·9236	9241	9246	9251	9255	9260	9265	9270	9275	9279	1	2	2	3	4
58	9·9284	9289	9294	9298	9303	9308	9312	9317	9322	9326	1	2	2	3	4
59	9·9331	9335	9340	9344	9349	9353	9358	9362	9367	9371	1	1	2	3	4.
60	9·9375	9380	9384	9388	9393	9397	9401	9406	9410	9414	1	1	2	3	4.
61	9·9418	9422	9427	9431	9435	9439	9443	9447	9451	9455	1	1	2	3	3
62	9·9459	9463	9467	9471	9475	9479	9483	9487	9491	9495	1	1	2	3	3
63	9·9499	9503	9507	9510	9514	9518	9522	9525	9529	9533	1	1	2	3	3
64	9·9537	9540	9544	9548	9551	9555	9558	9562	9566	9569	1	1	2	2	3
65	9·9573	9576	9580	9583	9587	9590	9594	9597	9601	9604	1	1	2	2	3
66	9·9607	9611	9614	9617	9621	9624	9627	9631	9634	9637	1	1	2	2	3
67	9·9640	9643	9647	9650	9653	9656	9659	9662	9666	9669	1	1	2	2	3
68	9·9672	9675	9678	9681	9684	9687	9690	9693	9696	9699	0	1	1	2	2
69	9·9702	9704	9707	9710	9713	9716	9719	9722	9724	9727	0	1	1	2	2
70	9·9730	9733	9735	9738	9741	9743	9746	9749	9751	9754	0	1	1	2	2
71	9·9757	9759	9762	9764	9767	9770	9772	9775	9777	9780	0	1	1	2	2
72	9·9782	9785	9787	9789	9792	9794	9797	9799	9801	9804	0	1	1	2	2
73	9·9806	9808	9811	9813	9815	9817	9820	9822	9824	9826	0	1	1	2	2
74	9·9828	9831	9833	9835	9837	9839	9841	9843	9845	9847	0	1	1	1	2
75	9·9849	9851	9853	9855	9857	9859	9861	9863	9865	9867	0	1	1	1	2
76	9·9869	9871	9873	9875	9876	9878	9880	9882	9884	9885	0	1	1	1	1
77	9·9887	9889	9891	9892	9894	9896	9897	9899	9901	9902	0	1	1	1	1
78	9·9904	9906	9907	9909	9910	9912	9913	9915	9916	9918	0	1	1	1	1
79	9·9919	9921	9922	9924	9925	9927	9928	9929	9931	9932	0	0	1	1	1
80	9·9934	9935	9936	9937	9939	9940	9941	9943	9944	9945	0	0	1	1	1
81	9·9946	9947	9949	9950	9951	9952	9953	9954	9955	9956	0	0	1	1	1
82	9·9958	9959	9960	9961	9962	9963	9964	9965	9966	9967	0	0	1	1	1
83	9·9968	9968	9969	9970	9971	9972	9973	9974	9975	9975	0	0	0	1	1
84	9·9976	9977	9978	9978	9979	9980	9981	9981	9982	9983	0	0	0	0	1
85	9·9983	9984	9985	9985	9986	9987	9987	9988	9988	9989	0	0	0	0	0
86	9·9989	9990	9990	9991	9991	9992	9992	9993	9993	9994	0	0	0	0	0
87	9·9994	9994	9995	9995	9996	9996	9996	9996	9997	9997	0	0	0	0	0
88	9·9997	9998	9998	9998	9998	9999	9999	9999	9999	9999	0	0	0	0	0
89	9·9999	9999	0̄000	0̄000	0̄000	0̄000	0̄000	0̄000	0̄000	0̄000	0	0	0	0	0

	0′	6′	12′	18′	24′	30′	36′	42′	48′	54′	1	2	3	4	5
0°	10·0000	0000	0000	0000	0000	0000	0000	0000	0000	9·9999	0	0	0	0	0
1	9·9999	9999	9999	9999	9999	9999	9998	9998	9998	9998	0	0	0	0	0
2	9·9997	9997	9997	9996	9996	9996	9996	9995	9995	9994	0	0	0	0	0
3	9·9994	9994	9993	9993	9992	9992	9991	9991	9990	9990	0	0	0	0	0
4	9·9989	9989	9988	9988	9987	9987	9986	9985	9985	9984	0	0	0	0	0
5	9·9983	9983	9982	9981	9981	9980	9979	9978	9978	9977	0	0	0	0	1
6	9·9976	9975	9975	9974	9973	9972	9971	9970	9969	9968	0	0	0	1	1
7	9·9968	9967	9966	9965	9964	9963	9962	9961	9960	9959	0	0	1	1	1
8	9·9957	9956	9955	9954	9953	9952	9951	9950	9949	9947	0	0	1	1	1
9	9·9946	9945	9944	9943	9941	9940	9939	9937	9936	9935	0	0	1	1	1
10	9·9934	9932	9931	9929	9928	9927	9925	9924	9922	9921	0	0	1	1	1
11	9·9919	9918	9916	9915	9913	9912	9910	9909	9907	9906	0	1	1	1	1
12	9·9904	9902	9901	9899	9897	9896	9894	9892	9891	9889	0	1	1	1	1
13	9·9887	9885	9884	9882	9880	9878	9876	9875	9873	9871	0	1	1	1	2
14	9·9869	9867	9865	9863	9861	9859	9857	9855	9853	9851	0	1	1	1	2
15	9·9849	9847	9845	9843	9841	9839	9837	9835	9833	9831	0	1	1	1	2
16	9·9828	9826	9824	9822	9820	9817	9815	9813	9811	9808	0	1	1	2	2
17	9·9806	9804	9801	9799	9797	9794	9792	9789	9787	9785	0	1	1	2	2
18	9·9782	9780	9777	9775	9772	9770	9767	9764	9762	9759	0	1	1	2	2
19	9·9757	9754	9751	9749	9746	9743	9741	9738	9735	9733	0	1	1	2	2
20	9·9730	9727	9724	9722	9719	9716	9713	9710	9707	9704	0	1	1	2	2
21	9·9702	9699	9696	9693	9690	9687	9684	9681	9678	9675	0	1	1	2	2
22	9·9672	9669	9666	9662	9659	9656	9653	9650	9647	9643	1	1	2	2	3
23	9·9640	9637	9634	9631	9627	9624	9621	9617	9614	9611	1	1	2	2	3
24	9·9607	9604	9601	9597	9594	9590	9587	9583	9580	9576	1	1	2	2	3
25	9·9573	9569	9566	9562	9558	9555	9551	9548	9544	9540	1	1	2	2	3
26	9·9537	9533	9529	9525	9522	9518	9514	9510	9507	9503	1	1	2	3	3
27	9·9499	9495	9491	9487	9483	9479	9475	9471	9467	9463	1	1	2	3	3
28	9·9459	9455	9451	9447	9443	9439	9435	9431	9427	9422	1	1	2	3	3
29	9·9418	9414	9410	9406	9401	9397	9393	9388	9384	9380	1	1	2	3	4
30	9·9375	9371	9367	9362	9358	9353	9349	9344	9340	9335	1	1	2	3	4
31	9·9331	9326	9322	9317	9312	9308	9303	9298	9294	9289	1	2	2	3	4
32	9·9284	9279	9275	9270	9265	9260	9255	9251	9246	9241	1	2	2	3	4
33	9·9236	9231	9226	9221	9216	9211	9206	9201	9196	9191	1	2	3	3	4
34	9·9186	9181	9175	9170	9165	9160	9155	9149	9144	9139	1	2	3	3	4
35	9·9134	9128	9123	9118	9112	9107	9101	9096	9091	9085	1	2	3	4	5
36	9·9080	9074	9069	9063	9057	9052	9046	9041	9035	9029	1	2	3	4	5
37	9·9023	9018	9012	9006	9000	8995	8989	8983	8977	8971	1	2	3	4	5
38	9·8965	8959	8953	8947	8941	8935	8929	8923	8917	8911	1	2	3	4	5
39	9·8905	8899	8893	8887	8880	8874	8868	8862	8855	8849	1	2	3	4	5
40	9·8843	8836	8830	8823	8817	8810	8804	8797	8791	8784	1	2	3	4	5
41	9·8778	8771	8765	8758	8751	8745	8738	8731	8724	8718	1	2	3	5	6
42	9·8711	8704	8697	8690	8683	8676	8669	8662	8655	8648	1	2	3	5	6
43	9·8641	8634	8627	8620	8613	8606	8598	8591	8584	8577	1	2	4	5	6
44	9·8569	8562	8555	8547	8540	8532	8525	8517	8510	8502	1	2	4	5	6

N.B.—Numbers in difference columns to be subtracted, not added.—*See Rules.*

	0'	6'	12'	18'	24'	30'	36'	42'	48'	54'	1 2 3	4 5
45°	9·8495	8487	8480	8472	8464	8457	8449	8441	8433	8426	1 3 4	5 6
46	9·8418	8410	8402	8394	8386	8378	8370	8362	8354	8346	1 3 4	5 7
47	9·8338	8330	8322	8313	8305	8297	8289	8280	8272	8264	1 3 4	6 7
48	9·8255	8247	8238	8230	8221	8213	8204	8195	8187	8178	1 3 4	6 7
49	9·8169	8161	8152	8143	8134	8125	8117	8108	8099	8090	1 3 4	6 7
50	9·8081	8072	8063	8053	8044	8035	8026	8017	8007	7998	2 3 5	6 8
51	9·7989	7979	7970	7960	7951	7941	7932	7922	7913	7903	2 3 5	6 8
52	9·7893	7884	7874	7864	7854	7844	7835	7825	7815	7805	2 3 5	7 8
53	9·7795	7785	7774	7764	7754	7744	7734	7723	7713	7703	2 3 5	7 9
54	9·7692	7682	7671	7661	7650	7640	7629	7618	7607	7597	2 4 5	7 9
55	9·7586	7575	7564	7553	7542	7531	7520	7509	7498	7487	2 4 6	7 9
56	9·7476	7464	7453	7442	7430	7419	7407	7396	7384	7373	2 4 6	8 10
57	9·7361	7349	7338	7326	7314	7302	7290	7278	7266	7254	2 4 6	8 10
58	9·7242	7230	7218	7205	7193	7181	7168	7156	7144	7131	2 4 6	8 10
59	9·7118	7106	7093	7080	7068	7055	7042	7029	7016	7003	2 4 6	9 11
60	9·6990	6977	6963	6950	6937	6923	6910	6896	6883	6869	2 4 7	9 11
61	9·6856	6842	6828	6814	6801	6787	6773	6759	6744	6730	2 5 7	9 12
62	9·6716	6702	6687	6673	6659	6644	6629	6615	6600	6585	2 5 7	10 12
63	9·6570	6556	6541	6526	6510	6495	6480	6465	6449	6434	3 5 8	10 13
64	9·6418	6403	6387	6371	6356	6340	6324	6308	6292	6276	3 5 8	11 13
65	9·6259	6243	6227	6210	6194	6177	6161	6144	6127	6110	3 6 8	11 14
66	9·6093	6076	6059	6042	6024	6007	5990	5972	5954	5937	3 6 9	12 15
67	9·5919	5901	5883	5865	5847	5828	5810	5792	5773	5754	3 6 9	12 15
68	9·5736	5717	5698	5679	5660	5641	5621	5602	5583	5563	3 6 10	13 16
69	9·5543	5523	5504	5484	5463	5443	5423	5402	5382	5361	3 7 10	14 17
70	9·5341	5320	5299	5278	5256	5235	5213	5192	5170	5148	4 7 11	14 18
71	9·5126	5104	5082	5060	5037	5015	4992	4969	4946	4923	4 8 11	15 19
72	9·4900	4876	4853	4829	4805	4781	4757	4733	4709	4684	4 8 12	16 20
73	9·4659	4634	4609	4584	4559	4533	4508	4482	4456	4430	4 9 13	17 21
74	9·4403	4377	4350	4323	4296	4269	4242	4214	4186	4158	5 9 14	18 23
75	9·4130	4102	4073	4044	4015	3986	3957	3927	3897	3867	5 10 15	20 24
76	9·3837	3806	3775	3745	3713	3682	3650	3618	3586	3554	5 11 16	21 26
77	9·3521	3488	3455	3421	3387	3353	3319	3284	3250	3214	6 11 17	23 28
78	9·3179	3143	3107	3070	3034	2997	2959	2921	2883	2845	6 12 19	25 31
79	9·2806	2767	2727	2687	2647	2606	2565	2524	2482	2439	7 14 20	27 34
80	9·2397	2353	2310	2266	2221	2176	2131	2085	2038	1991	8 15 23	30 38
81	9·1943	1895	1847	1797	1747	1697	1646	1594	1542	1489	8 17 25	34 42
82	9·1436	1381	1326	1271	1214	1157	1099	1040	0981	0920	10 19 29	38 48
83	9·0859	0797	0734	0670	0605	0539	0472	0403	0334	0264	11 22 33	44 55
84	9·0192	0120	0046	9970	9894	9816	9736	9655	9573	9489	13 26 39	52 65
85	8·9403	9315	9226	9135	9042	8946	8849	8749	8647	8543	16 32 48	64 80
86	8·8436	8326	8213	8098	7979	7857	7731	7602	7468	7330	21 41 62	82 103
87	8·7188	7041	6889	6731	6567	6397	6220	6035	5842	5640		
88	8·5428	5206	4971	4723	4459	4179	3880	3558	3210	2832		
89	8·2419	1961	1450	0870	0200	7·9408	8439	7190	5429	2419		

N.B.—Numbers in difference columns to be subtracted, not added.—*See Rules.*

	0′	6′	12′	18′	24′	30′	36′	42′	48′	54′	1	2	3	4	5
0°	Inf. Neg.	7·2419	5429	7190	8439	9409	0̄200	0̄870	1̄450	1̄962					
1	8·2419	2833	3211	3559	3881	4181	4461	4725	4973	5208					
2	8·5431	5643	5845	6038	6223	6401	6571	6736	6894	7046	29	58	87	116	145
3	8·7194	7337	7475	7609	7739	7865	7988	8107	8223	8336	21	41	62	83	103
4	8·8446	8554	8659	8762	8862	8960	9056	9150	9241	9331	16	32	48	64	81
5	8·9420	9506	9591	9674	9756	9836	9915	9992	0̄068	0̄143	13	26	40	53	66
6	9·0216	0289	0360	0430	0499	0567	0633	0699	0764	0828	11	22	34	45	56
7	9·0891	0954	1015	1076	1135	1194	1252	1310	1367	1423	10	20	29	39	49
8	9·1478	1533	1587	1640	1693	1745	1797	1848	1898	1948	9	17	26	35	43
9	9·1997	2046	2094	2142	2189	2236	2282	2328	2374	2419	8	16	23	31	39
10	9·2463	2507	2551	2594	2637	2680	2722	2764	2805	2846	7	14	21	28	35
11	9·2887	2927	2967	3006	3046	3085	3123	3162	3200	3237	6	13	19	26	32
12	9·3275	3312	3349	3385	3422	3458	3493	3529	3564	3599	6	12	18	24	30
13	9·3634	3668	3702	3736	3770	3804	3837	3870	3903	3935	6	11	17	22	28
14	9·3968	4000	4032	4064	4095	4127	4158	4189	4220	4250	5	10	16	21	26
15	9·4281	4311	4341	4371	4400	4430	4459	4488	4517	4546	5	10	15	20	25
16	9·4575	4603	4632	4660	4688	4716	4744	4771	4799	4826	5	9	14	19	23
17	9·4853	4880	4907	4934	4961	4987	5014	5040	5066	5092	4	9	13	18	22
18	9·5118	5143	5169	5195	5220	5245	5270	5295	5320	5345	4	8	13	17	21
19	9·5370	5394	5419	5443	5467	5491	5516	5539	5563	5587	4	8	12	16	20
20	9·5611	5634	5658	5681	5704	5727	5750	5773	5796	5819	4	8	12	15	19
21	9·5842	5864	5887	5909	5932	5954	5976	5998	6020	6042	4	7	11	15	19
22	9·6064	6086	6108	6129	6151	6172	6194	6215	6236	6257	4	7	11	14	18
23	9·6279	6300	6321	6341	6362	6383	6404	6424	6445	6465	3	7	10	14	17
24	9·6486	6506	6527	6547	6567	6587	6607	6627	6647	6667	3	7	10	13	17
25	9·6687	6706	6726	6746	6765	6785	6804	6824	6843	6863	3	7	10	13	16
26	9·6882	6901	6920	6939	6958	6977	6996	7015	7034	7053	3	6	9	13	16
27	9·7072	7090	7109	7128	7146	7165	7183	7202	7220	7238	3	6	9	12	15
28	9·7257	7275	7293	7311	7330	7348	7366	7384	7402	7420	3	6	9	12	15
29	9·7438	7455	7473	7491	7509	7526	7544	7562	7579	7597	3	6	9	12	15
30	9·7614	7632	7649	7667	7684	7701	7719	7736	7753	7771	3	6	9	12	14
31	9·7788	7805	7822	7839	7856	7873	7890	7907	7924	7941	3	6	9	11	14
32	9·7958	7975	7992	8008	8025	8042	8059	8075	8092	8109	3	6	8	11	14
33	9·8125	8142	8158	8175	8191	8208	8224	8241	8257	8274	3	5	8	11	14
34	9·8290	8306	8323	8339	8355	8371	8388	8404	8420	8436	3	5	8	11	14
35	9·8452	8468	8484	8501	8517	8533	8549	8565	8581	8597	3	5	8	11	13
36	9·8613	8629	8644	8660	8676	8692	8708	8724	8740	8755	3	5	8	11	13
37	9·8771	8787	8803	8818	8834	8850	8865	8881	8897	8912	3	5	8	10	13
38	9·8928	8944	8959	8975	8990	9006	9022	9037	9053	9068	3	5	8	10	13
39	9·9084	9099	9115	9130	9146	9161	9176	9192	9207	9223	3	5	8	10	13
40	9·9238	9254	9269	9284	9300	9315	9330	9346	9361	9376	3	5	8	10	13
41	9·9392	9407	9422	9438	9453	9468	9483	9499	9514	9529	3	5	8	10	13
42	9·9544	9560	9575	9590	9605	9621	9636	9651	9666	9681	3	5	8	10	13
43	9·9697	9712	9727	9742	9757	9773	9788	9803	9818	9833	3	5	8	10	13
44	9·9848	9864	9879	9894	9909	9924	9939	9955	9970	9985	3	5	8	10	13

	0′	6′	12′	18′	24′	30′	36′	42′	48′	54′	1 2
45°	10·0000	0015	0030	0045	0061	0076	0091	0106	0121	0136	3 5
46	10·0152	0167	0182	0197	0212	0228	0243	0258	0273	0288	3 5
47	10·0303	0319	0334	0349	0364	0379	0395	0410	0425	0440	3 5
48	10·0456	0471	0486	0501	0517	0532	0547	0562	0578	0593	3 5
49	10·0608	0624	0639	0654	0670	0685	0700	0716	0731	0746	3 5
50	10·0762	0777	0793	0808	0824	0839	0854	0870	0885	0901	3 5
51	10·0916	0932	0947	0963	0978	0994	1010	1025	1041	1056	3 5
52	10·1072	1088	1103	1119	1135	1150	1166	1182	1197	1213	3 5
53	10·1229	1245	1260	1276	1292	1308	1324	1340	1356	1371	3 5
54	10·1387	1403	1419	1435	1451	1467	1483	1499	1516	1532	3 5
55	10·1548	1564	1580	1596	1612	1629	1645	1661	1677	1694	3 5
56	10·1710	1726	1743	1759	1776	1792	1809	1825	1842	1858	3 5
57	10·1875	1891	1908	1925	1941	1958	1975	1992	2008	2025	3 6
58	10·2042	2059	2076	2093	2110	2127	2144	2161	2178	2195	3 6
59	10·2212	2229	2247	2264	2281	2299	2316	2333	2351	2368	3 6
60	10·2386	2403	2421	2438	2456	2474	2491	2509	2527	2545	3 6
61	10·2562	2580	2598	2616	2634	2652	2670	2689	2707	2725	3 6
62	10·2743	2762	2780	2798	2817	2835	2854	2872	2891	2910	3 6
63	10·2928	2947	2966	2985	3004	3023	3042	3061	3080	3099	3 6
64	10·3118	3137	3157	3176	3196	3215	3235	3254	3274	3294	3 6
65	10·3313	3333	3353	3373	3393	3413	3433	3453	3473	3494	3 7
66	10·3514	3535	3555	3576	3596	3617	3638	3659	3679	3700	3 7
67	10·3721	3743	3764	3785	3806	3828	3849	3871	3892	3914	4 .7
68	10·3936	3958	3980	4002	4024	4046	4068	4091	4113	4136	4 7
69	10·4158	4181	4204	4227	4250	4273	4296	4319	4342	4366	4 8
70	10·4389	4413	4437	4461	4484	4509	4533	4557	4581	4606	4 8
71	10·4630	4655	4680	4705	4730	4755	4780	4805	4831	4857	4 8
72	10·4882	4908	4934	4960	4986	5013	5039	5066	5093	5120	4 9
73	10·5147	5174	5201	5229	5256	5284	5312	5340	5368	5397	5 9
74	10·5425	5454	5483	5512	5541	5570	5600	5629	5659	5689	5 10
75	10·5719	5750	5780	5811	5842	5873	5905	5936	5968	6000	5 10
76	10·6032	6065	6097	6130	6163	6196	6230	6264	6298	6332	6 11
77	10·6366	6401	6436	6471	6507	6542	6578	6615	6651	6688	6 12
78	10·6725	6763	6800	6838	6877	6915	6954	6994	7033	7073	6 13
79	10·7113	7154	7195	7236	7278	7320	7363	7406	7449	7493	7 14
80	10·7537	7581	7626	7672	7718	7764	7811	7858	7906	7954	8 16
81	10·8003	8052	8102	8152	8203	8255	8307	8360	8413	8467	9 17
82	10·8522	8577	8633	8690	8748	8806	8865	8924	8985	9046	10 20
83	10·9109	9172	9236	9301	9367	9433	9501	9570	9640	9711	11 22
84	10·9784	9857	9932	ō008	ō085	ō164	ō244	ō326	ō409	ō494	13 26
85	11·0580	0669	0759	0850	0944	1040	1138	1238	1341	1446	16 32
86	11·1554	1664	1777	1893	2012	2135	2261	2391	2525	2663	20 41
87	11·2806	2954	3106	3264	3429	3599	3777	3962	4155	4357	29 58
88	11·4569	4792	5027	5275	5539	5819	6119	6441	6789	7167	
89	11·7581	8038	8550	9130	9800	ō591	ī561	2̄810	4̄571	7581	

LOGARITHMIC COTANGENTS.

O'	6'	12'	18'	24'	30'	36'	42'	48'	54'	1	2	3	4	5
Inf.	12·76	4571	2810	1561	0591	9800	9130	8550	8038					
11·7581	7167	6789	6441	6119	5819	5539	5275	5027	4792					
11·4569	4357	4155	3962	3777	3599	3429	3264	3106	2954	29	59	88	118	147
11·2806	2663	2525	2391	2261	2135	2012	1893	1777	1664	21	42	63	83	104
11·1554	1446	1341	1238	1138	1040	0944	0850	0759	0669	16	32	49	65	81
11·0580	0494	0409	0326	0244	0164	0085	0008	9932	9857	13	26	40	53	66
10·9784	9711	9640	9570	9501	9433	9367	9301	9236	9172	11	23	34	45	55
10·9109	9046	8985	8924	8865	8806	8748	8690	8633	8577	10	20	29	39	49
10·8522	8467	8413	8360	8307	8255	8203	8152	8102	8052	9	17	26	35	43
10·8003	7954	7906	7858	7811	7764	7718	7672	7626	7581	8	16	23	31	39
10·7537	7493	7449	7406	7363	7320	7278	7236	7195	7154	7	14	21	28	35
10·7113	7073	7033	6994	6954	6915	6877	6838	6800	6763	6	13	19	26	32
10·6725	6688	6651	6615	6578	6542	6507	6471	6436	6401	6	12	18	24	30
10·6366	6332	6298	6264	6230	6196	6163	6130	6097	6065	6	11	17	22	28
10·6032	6000	5968	5936	5905	5873	5842	5811	5780	5750	5	10	15	21	26
10·5719	5689	5659	5629	5600	5570	5541	5512	5483	5454	5	10	15	20	25
10·5425	5397	5368	5340	5312	5284	5256	5229	5201	5174	5	9	14	19	23
10·5147	5120	5093	5066	5039	5013	4986	4960	4934	4908	4	9	13	18	22
10·4882	4857	4831	4805	4780	4755	4730	4705	4680	4655	4	9	13	17	21
10·4630	4606	4581	4557	4533	4509	4484	4461	4437	4413	4	8	12	16	20
10·4389	4366	4342	4319	4296	4273	4250	4227	4204	4181	4	8	12	15	19
10·4158	4136	4113	4091	4068	4046	4024	4002	3980	3958	4	7	11	15	19
10·3936	3914	3892	3871	3849	3828	3806	3785	3764	3743	4	7	11	14	18
10·3721	3700	3679	3659	3638	3617	3596	3576	3555	3535	3	7	10	14	17
10·3514	3494	3473	3453	3433	3413	3393	3373	3353	3333	3	7	10	13	17
10·3313	3294	3274	3254	3235	3215	3196	3176	3157	3137	3	6	10	13	16
10·3118	3099	3080	3061	3042	3023	3004	2985	2966	2947	3	6	9	13	16
10·2928	2910	2891	2872	2854	2835	2817	2798	2780	2762	3	6	9	12	15
10·2743	2725	2707	2689	2670	2652	2634	2616	2598	2580	3	6	9	12	15
10·2562	2545	2527	2509	2491	2474	2456	2438	2421	2403	3	6	9	12	15
10·2386	2368	2351	2333	2316	2299	2281	2264	2247	2229	3	6	9	12	14
10·2212	2195	2178	2161	2144	2127	2110	2093	2076	2059	3	6	9	11	14
10·2042	2025	2008	1992	1975	1958	1941	1925	1908	1891	3	5	8	11	14
10·1875	1858	1842	1825	1809	1792	1776	1759	1743	1726	3	5	8	11	14
10·1710	1694	1677	1661	1645	1629	1612	1596	1580	1564	3	5	8	11	14
10·1548	1532	1516	1499	1483	1467	1451	1435	1419	1403	3	5	8	11	13
10·1387	1371	1356	1340	1324	1308	1292	1276	1260	1245	3	5	8	11	13
10·1229	1213	1197	1182	1166	1150	1135	1119	1103	1088	3	5	8	10	13
10·1072	1056	1041	1025	1010	0994	0978	0963	0947	0932	3	5	8	10	13
10·0916	0901	0885	0870	0854	0839	0824	0808	0793	0777	3.	5	8	10	13
10·0762	0746	0731	0716	0700	0685	0670	0654	0639	0624	3	5	8	10	13
10·0608	0593	0578	0562	0547	0532	0517	0501	0486	0471	3	5	8	10	13
10·0456	0440	0425	0410	0395	0379	0364	0349	0334	0319	3	5	8	10	13
10·0303	0288	0273	0258	0243	0228	0212	0197	0182	0167	3.	5	8	10	13
10·0152	0136	0121	0106	0091	0076	0061	0045	0030	0015	3	5	8	10	13

N.B.—Numbers in difference columns to be subtracted, not added.—*See Rules.*

	0'	6'	12'	18'	24'	30'	36'	42'	48'	54'	1	2	3	4	5
45°	10·0000	$\overline{9985}$	$\overline{9970}$	$\overline{9955}$	$\overline{9939}$	$\overline{9924}$	$\overline{9909}$	$\overline{9894}$	$\overline{9879}$	$\overline{9864}$	3	5	8	10	13
46	9·9848	9833	9818	9803	9788	9773	9757	9742	9727	9712	3	5	8	10	13
47	9·9697	9681	9666	9651	9636	9621	9605	9590	9575	9560	3	5	8	10	13
48	9·9544	9529	9514	9499	9483	9468	9453	9438	9422	9407	3	5	8	10	13
49	9·9392	9376	9361	9346	9330	9315	9300	9284	9269	9254	3	5	8	10	13
50	9·9238	9223	9207	9192	9176	9161	9146	9130	9115	9099	3	5	8	10	13
51	9·9084	9068	9053	9037	9022	9006	8990	8975	8959	8944	3	5	8	10	13
52	9·8928	8912	8897	8881	8865	8850	8834	8818	8803	8787	3	5	8	10	13
53	9·8771	8755	8740	8724	8708	8692	8676	8660	8644	8629	3	5	8	11	13
54	9·8613	8597	8581	8565	8549	8533	8517	8501	8484	8468	3	5	8	11	13
55	9·8452	8436	8420	8404	8388	8371	8355	8339	8323	8306	3	5	8	11	14
56	9·8290	8274	8257	8241	8224	8208	8191	8175	8158	8142	3	5	8	11	14
57	9·8125	8109	8092	8075	8059	8042	8025	8008	7992	7975	3	6	8	11	14
58	9·7958	7941	7924	7907	7890	7873	7856	7839	7822	7805	3	6	9	11	14
59	9·7788	7771	7753	7736	7719	7701	7684	7667	7649	7632	3	6	9	12	14
60	9·7614	7597	7579	7562	7544	7526	7509	7491	7473	7455	3	6	9	12	15
61	9·7438	7420	7402	7384	7366	7348	7330	7311	7293	7275	3	6	9	12	15
62	9·7257	7238	7220	7202	7183	7165	7146	7128	7109	7090	3	6	9	12	15
63	9·7072	7053	7034	7015	6996	6977	6958	6939	6920	6901	3	6	9	13	16
64	9·6882	6863	6843	6824	6804	6785	6765	6746	6726	6706	3	7	10	13	16
65	9·6687	6667	6647	6627	6607	6587	6567	6547	6527	6506	3	7	10	13	17
66	9·6486	6465	6445	6424	6404	6383	6362	6341	6321	6300	3	7	10	14	17
67	9·6279	6257	6236	6215	6194	6172	6151	6129	6108	6086	4	7	11	14	18
68	9·6064	6042	6020	5998	5976	5954	5932	5909	5887	5864	4	7	11	15	19
69	9·5842	5819	5796	5773	5750	5727	5704	5681	5658	5634	4	8	12	15	19
70	9·5611	5587	5563	5539	5516	5491	5467	5443	5419	5394	4	8	12	16	20
71	9·5370	5345	5320	5295	5270	5245	5220	5195	5169	5143	4	8	13	17	21
72	9·5118	5092	5066	5040	5014·	4987	4961	4934	4907	4880	4	9	13	18	22
73	9·4853	4826	4799	4771	4744	4716	4688	4660	4632	4603	5	9	14	19	23
74	9·4575	4546	4517	4488	4459	4430	4400	4371	4341	4311	5	10	15	20	25
75	9·4281	4250	4220	4189	4158	4127	4095	4064	4032	4000	5	10	16	21	26
76	9·3968	3935	3903	3870	3837	3804	3770	3736	3702	3668	6	11	17	22	28
77	9·3634	3599	3564	3529	3493	3458	3422	3385	3349	3312	6	12	18	24	30
78	9·3275	3237	3200	3162	3123	3085	3046	3006	2967	2927	6	13	19	26	32
79	9·2887	2846	2805	2764	2722	2680	2637	2594	2551	2507	7	14	21	28	35
80	9·2463	2419	2374	2328	2282	2236	2189	2142	2094	2046	8	16	23	31	39
81	9·1997	1948	1898	1848	1797	1745	1693	1640	1587	1533	9	17	26	35	43
82	9·1478	1423	1367	1310	1252	1194	1135	1076	1015	0954	10	20	29	39	49
83	9·0891	0828	0764	0699	0633	0567	0499	0430	0360	0289	11	23	34	45	56
84	9·0216	0143	0068	9992	9915	9836	9756	9674	9591	9506	13	27	40	53	66
85	8·9420	9331	9241	9150	9056	8960	8862	8762	8659	8554	16	32	49	65	81
86	8·8446	8336	8223	8107	7988	7865	7739	7609	7475	7337	21	42	63	83	104
87	8·7194	7046	6894	6736	6571	6401	6223	6038	5845	5643	29	59	88	118	147
88	8·5431	5208	4973	4725	4461	4181	3881	3559	3211	2833					
89	8·2419	1962	1450	0870	0200	$\overline{9409}$	$\overline{8439}$	$\overline{7190}$	$\overline{5429}$	$\overline{2419}$					

N.B.—Numbers in difference columns to be subtracted, not added.—*See Rules.*

	0′	6′	12′	18′	24′	30′	36′	42′	48′	54′	1	2	3	4	5
0°	10·0000	0000	0000	0000	0000	0000	0000	0000	0000	0001	0	0	0	0	0
1	10·0001	0001	0001	0001	0001	0001	0002	0002	0002	0002	0	0	0	0	0
2	10·0003	0003	0003	0004	0004	0004	0004	0005	0005	0006	0	0	0	0	0
3	10·0006	0006	0007	0007	0008	0008	0009	0009	0010	0010	0	0	0	0	0
4	10·0011	0011	0012	0012	0013	0013	0014	0015	0015	0016	0	0	0	0	0
5	10·0017	0017	0018	0019	0019	0020	0021	0022	0022	0023	0	0	0	0	1
6	10·0024	0025	0025	0026	0027	0028	0029	0030	0031	0032	0	0	0	1	1
7	10·0032	0033	0034	0035	0036	0037	0038	0039	0040	0041	0	0	0	1	1
8	10·0042	0044	0045	0046	0047	0048	0049	0050	0051	0053	0	0	1	1	1
9	10·0054	0055	0056	0057	0059	0060	0061	0063	0064	0065	0	0	1	1	1
10	10·0066	0068	0069	0071	0072	0073	0075	0076	0078	0079	0	0	1	1	1
11	10·0081	0082	0084	0085	0087	0088	0090	0091	0093	0094	0	1	1	1	1
12	10·0096	0098	0099	0101	0103	0104	0106	0108	0109	0111	0	1	1	1	1
13	10·0113	0115	0116	0118	0120	0122	0124	0125	0127	0129	0	1	1	1	2
14	10·0131	0133	0135	0137	0139	0141	0143	0145	0147	0149	0	1	1	1	2
15	10·0151	0153	0155	0157	0159	0161	0163	0165	0167	0169	0	1	1	1	2
16	10·0172	0174	0176	0178	0180	0183	0185	0187	0189	0192	0	1	1	1	2
17	10·0194	0196	0199	0201	0203	0206	0208	0211	0213	0215	0	1	1	2	2
18	10·0218	0220	0223	0225	0228	0230	0233	0236	0238	0241	0	1	1	2	2
19	10·0243	0246	0249	0251	0254	0257	0259	0262	0265	0267	0	1	1	2	2
20	10·0270	0273	0276	0278	0281	0284	0287	0290	0293	0296	0	1	1	2	2
21	10·0298	0301	0304	0307	0310	0313	0316	0319	0322	0325	0	1	1	2	2
22	10·0328	0331	0334	0338	0341	0344	0347	0350	0353	0357	1	1	2	2	3
23	10·0360	0363	0366	0369	0373	0376	0379	0383	0386	0389	1	1	2	2	3
24	10·0393	0396	0399	0403	0406	0410	0413	0417	0420	0424	1	1	2	2	3
25	10·0427	0431	0434	0438	0442	0445	0449	0452	0456	0460	1	1	2	2	3
26	10·0463	0467	0471	0475	0478	0482	0486	0490	0494	0497	1	1	2	3	3
27	10·0501	0505	0509	0513	0517	0521	0525	0529	0533	0537	1	1	2	3	3
28	10·0541	0545	0549	0553	0557	0561	0565	0569	0573	0578	1	1	2	3	3
29	10·0582	0586	0590	0594	0599	0603	0607	0612	0616	0620	1	1	2	3	4
30	10·0625	0629	0633	0638	0642	0647	0651	0656	0660	0665	1	1	2	3	4
31	10·0669	0674	0678	0683	0688	0692	0697	0702	0706	0711	1	2	2	3	4
32	10·0716	0721	0725	0730	0735	0740	0745	0749	0754	0759	1	2	2	3	4
33	10·0764	0769	0774	0779	0784	0789	0794	0799	0804	0809	1	2	2	3	4
34	10·0814	0819	0825	0830	0835	0840	0845	0851	0856	0861	1	2	3	3	4
35	10·0866	0872	0877	0882	0888	0893	0899	0904	0909	0915	1	2	3	4	5
36	10·0920	0926	0931	0937	0943	0948	0954	0959	0965	0971	1	2	3	4	5
37	10·0977	0982	0988	0994	1000	1005	1011	1017	1023	1029	1	2	3	4	5
38	10·1035	1041	1047	1053	1059	1065	1071	1077	1083	1089	1	2	3	4	5
39	10·1095	1101	1107	1113	1120	1126	1132	1138	1145	1151	1	2	3	4	5
40	10·1157	1164	1170	1177	1183	1190	1196	1203	1209	1216	1	2	3	4	5
41	10·1222	1229	1235	1242	1249	1255	1262	1269	1276	1282	1	2	3	4	6
42	10·1289	1296	1303	1310	1317	1324	1331	1338	1345	1352	1	2	3	5	6
43	10·1359	1366	1373	1380	1387	1394	1402	1409	1416	1423	1	2	4	5	6
44	10·1431	1438	1445	1453	1460	1468	1475	1483	1490	1498	1	2	4	5	6

LOGARITHMIC SECANTS.

	0'	6'	12'	18'	24'	30'	36'	42'	48'
45°	10·1505	1513	1520	1528	1536	1543	1551	1559	1567
46	10·1582	1590	1598	1606	1614	1622	1630	1638	1646
47	10·1662	1670	1678	1687	1695	1703	1711	1720	1728
48	10·1745	1753	1762	1770	1779	1787	1796	1805	1813
49	10·1831	1839	1848	1857	1866	1875	1883	1892	1901
50	10·1919	1928	1937	1947	1956	1965	1974	1983	1993
51	10·2011	2021	2030	2040	2049	2059	2068	2078	2087
52	10·2107	2116	2126	2136	2146	2156	2165	2175	2185
53	10·2205	2215	2226	2236	2246	2256	2266	2277	2287
54	10·2308	2318	2329	2339	2350	2360	2371	2382	2393
55	10·2414	2425	2436	2447	2458	2469	2480	2491	2502
56	10·2524	2536	2547	2558	2570	2581	2593	2604	2616
57	10·2639	2651	2662	2674	2686	2698	2710	2722	2734
58	10·2758	2770	2782	2795	2807	2819	2832	2844	2856
59	10·2882	2894	2907	2920	2932	2945	2958	2971	2984
60	10·3010	3023	3037	3050	3063	3077	3090	3104	3117
61	10·3144	3158	3172	3186	3199	3213	3227	3241	3256
62	10·3284	3298	3313	3327	3341	3356	3371	3385	3400
63	10·3430	3444	3459	3474	3490	3505	3520	3535	3551
64	10·3582	3597	3613	3629	3644	3660	3676	3692	3708
65	10·3741	3757	3773	3790	3806	3823	3839	3856	3873
66	10·3907	3924	3941	3958	3976	3993	4010	4028	4046
67	10·4081	4099	4117	4135	4153	4172	4190	4208	4227
68	10·4264	4283	4302	4321	4340	4359	4379	4398	4417
69	10·4457	4477	4496	4516	4537	4557	4577	4598	4618
70	10·4659	4680	4701	4722	4744	4765	4787	4808	4830
71	10·4874	4896	4918	4940	4963	4985	5008	5031	5054
72	10·5100	5124	5147	5171	5195	5219	5243	5267	5291
73	10·5341	5366	5391	5416	5441	5467	5492	5518	5544
74	10·5597	5623	5650	5677	5704	5731	5758	5786	5814
75	10·5870	5898	5927	5956	5985	6014	6043	6073	6103
76	10·6163	6194	6225	6255	6287	6318	6350	6382	6414
77	10·6479	6512	6545	6579	6613	6647	6681	6716	6750
78	10·6821	6857	6893	6930	6966	7003	7041	7079	7117
79	10·7194	7233	7273	7313	7353	7394	7435	7476	7518
80	10·7603	7647	7690	7734	7779	7824	7869	7915	7962
81	10·8057	8105	8153	8203	8253	8303	8354	8406	8458
82	10·8564	8619	8674	8729	8786	8843	8901	8960	9019
83	10·9141	9203	9266	9330	9395	9461	9528	9597	9666
84	10·9808	9880	9954	$\overline{0}$030	$\overline{0}$106	$\overline{0}$184	$\overline{0}$264	$\overline{0}$345	$\overline{0}$427
85	11·0597	0685	0774	0865	0958	1054	1151	1251	1353
86	11·1564	1674	1787	1902	2021	2143	2269	2398	2532
87	11·2812	2959	3111	3269	3433	3603	3780	3965	4158
88	11·4572	4794	5029	5277	5541	5821	6120	6442	6790
89	11·7581	8039	8550	9130	9800	$\overline{0}$592	$\overline{1}$561	$\overline{2}$810	4571

	O'	6'	12'	18'	24'	30'	36'	42'	48'	54'	1	2	3	4	5
0°	Inf.	12·76	4571	2810	1561	0592	9̄800	9̄130	8̄550	8̄039					
1	11·7581	7168	6790	6442	6120	5821	5541	5277	5029	4794					
2	11·4572	4360	4158	3965	3780	3603	3433	3269	3111	2959					
3	11·2812	2670	2532	2398	2269	2143	2021	1902	1787	1674					
4	11·1564	1457	1353	1251	1151	1054	0958	0865	0774	0685	16	32	48	64	81
5	11·0597	0511	0427	0345	0264	0184	0106	0030	9954	9880	13	26	39	53	66
6	10·9808	9736	9666	9597	9528	9461	9395	9330	9266	9203	11	22	33	44	56
7	10·9141	9080	9019	8960	8901	8843	8786	8729	8674	8619	10	19	29	38	48
8	10·8564	8511	8458	8406	8354	8303	8253	8203	8153	8105	8	17	25	34	42
9	10·8057	8009	7962	7915	7869	7824	7779	7734	7690	7647	8	15	23	30	38
10	10·7603	7561	7518	7476	7435	7394	7353	7313	7273	7233	7	14	20	27	34
11	10·7194	7155	7117	7079	7041	7003	6966	6930	6893	6857	6	12	19	25	31
12	10·6821	6786	6750	6716	6681	6647	6613	6579	6545	6512	6	11	17	23	28
13	10·6479	6446	6414	6382	6350	6318	6287	6255	6225	6194	5	11	16	21	26
14	10·6163	6133	6103	6073	6043	6014	5985	5956	5927	5898	5	10	15	20	24
15	10·5870	5842	5814	5786	5758	5731	5704	5677	5650	5623	5	9	14	18	23
16	10·5597	5570	5544	5518	5492	5467	5441	5416	5391	5366	4	9	13	17	21
17	10·5341	5316	5291	5267	5243	5219	5195	5171	5147	5124	4	8	12	16	20
18	10·5100	5077	5054	5031	5008	4985	4963	4940	4918	4896	4	8	11	15	19
19	10·4874	4852	4830	4808	4787	4765	4744	4722	4701	4680	4	7	11	14	18
20	10·4659	4639	4618	4598	4577	4557	4537	4516	4496	4477	3	7	10	13	17
21	10·4457	4437	4417	4398	4379	4359	4340	4321	4302	4283	3	6	10	13	16
22	10·4264	4246	4227	4208	4190	4172	4153	4135	4117	4099	3	6	9	12	15
23	10·4081	4063	4046	4028	4010	3993	3976	3958	3941	3924	3	6	9	12	14
24	10·3907	3890	3873	3856	3839	3823	3806	3790	3773	3757	3	6	8	11	14
25	10·3741	3724	3708	3692	3676	3660	3644	3629	3613	3597	3	5	8	11	13
26	10·3582	3566	3551	3535	3520	3505	3490	3474	3459	3444	3	5	8	10	13
27	10·3430	3415	3400	3385	3371	3356	3341	3327	3313	3298	2	5	7	10	12
28	10·3284	3270	3256	3241	3227	3213	3199	3186	3172	3158	2	5	7	9	12
29	10·3144	3131	3117	3104	3090	3077	3063	3050	3037	3023	2	4	7	9	11
30	10·3010	2997	2984	2971	2958	2945	2932	2920	2907	2894	2	4	6	8	11
31	10·2882	2869	2856	2844	2832	2819	2807	2795	2782	2770	2	4	6	8	10
32	10·2758	2746	2734	2722	2710	2698	2686	2674	2662	2651	2	4	6	8	10
33	10·2639	2627	2616	2604	2593	2581	2570	2558	2547	2536	2	4	6	8	10
34	10·2524	2513	2502	2491	2480	2469	2458	2447	2436	2425	2	4	5	7	9
35	10·2414	2403	2393	2382	2371	2360	2350	2339	2329	2318	2	4	5	7	9
36	10·2308	2297	2287	2277	2266	2256	2246	2236	2226	2215	2	3	5	7	9
37	10·2205	2195	2185	2175	2165	2156	2146	2136	2126	2116	2	3	5	7	8
38	10·2107	2097	2087	2078	2068	2059	2049	2040	2030	2021	2	3	5	6	8
39	10·2011	2002	1993	1983	1974	1965	1956	1947	1937	1928	2	3	5	6	8
40	10·1919	1910	1901	1892	1883	1875	1866	1857	1848	1839	1	3	4	6	7
41	10·1831	1822	1813	1805	1796	1787	1779	1770	1762	1753	1	3	4	6	7
42	10·1745	1736	1728	1720	1711	1703	1695	1687	1678	1670	1	3	4	6	7
43	10·1662	1654	1646	1638	1630	1622	1614	1606	1598	1590	1	3	4	5	7
44	10·1582	1574	1567	1559	1551	1543	1536	1528	1520	1513	1	3	4	5	6

N.B.—Numbers in difference columns to be subtracted, not added.—*See Rules.*

LOGARITHMIC COSECANTS.

0′	6′	12′	18′	24′	30′	36′	42′	48′	54′
10·1505	1498	1490	1483	1475	1468	1460	1453	1445	1438
10·1431	1423	1416	1409	1402	1394	1387	1380	1373	1366
10·1359	1352	1345	1338	1331	1324	1317	1310	1303	1296
10·1289	1282	1276	1269	1262	1255	1249	1242	1235	1229
10·1222	1216	1209	1203	1196	1190	1183	1177	1170	1164
10·1157	1151	1145	1138	1132	1126	1120	1113	1107	1101
10·1095	1089	1083	1077	1071	1065	1059	1053	1047	1041
10·1035	1029	1023	1017	1011	1005	1000	0994	0988	0982
10·0977	0971	0965	0959	0954	0948	0943	0937	0931	0926
10·0920	0915	0909	0904	0899	0893	0888	0882	0877	0872
10·0866	0861	0856	0851	0845	0840	0835	0830	0825	0819
10·0814	0809	0804	0799	0794	0789	0784	0779	0774	0769
10·0764	0759	0754	0749	0745	0740	0735	0730	0725	0721
10·0716	0711	0706	0702	0697	0692	0688	0683	0678	0674
10·0669	0665	0660	0656	0651	0647	0642	0638	0633	0629
10·0625	0620	0616	0612	0607	0603	0599	0594	0590	0586
10·0582	0578	0573	0569	0565	0561	0557	0553	0549	0545
10·0541	0537	0533	0529	0525	0521	0517	0513	0509	0505
10·0501	0497	0494	0490	0486	0482	0478	0475	0471	0467
10·0463	0460	0456	0452	0449	0445	0442	0438	0434	0431
10·0427	0424	0420	0417	0413	0410	0406	0403	0399	0396
10·0393	0389	0386	0383	0379	0376	0373	0369	0366	0363
10·0360	0357	0353	0350	0347	0344	0341	0338	0334	0331
10·0328	0325	0322	0319	0316	0313	0310	0307	0304	0301
10·0298	0296	0293	0290	0287	0284	0281	0278	0276	0273
10·0270	0267	0265	0262	0259	0257	0254	0251	0249	0246
10·0243	0241	0238	0236	0233	0230	0228	0225	0223	0220
10·0218	0215	0213	0211	0208	0206	0203	0201	0199	0196
10·0194	0192	0189	0187	0185	0183	0180	0178	0176	0174
10·0172	0169	0167	0165	0163	0161	0159	0157	0155	0153
10·0151	0149	0147	0145	0143	0141	0139	0137	0135	0133
10·0131	0129	0127	0125	0124	0122	0120	0118	0116	0115
10·0113	0111	0109	0108	0106	0104	0103	0101	0099	0098
10·0096	0094	0093	0091	0090	0088	0087	0085	0084	0082
10·0081	0079	0078	0076	0075	0073	0072	0071	0069	0068
10·0066	0065	0064	0063	0061	0060	0059	0057	0056	0055
10·0054	0053	0051	0050	0049	0048	0047	0046	0045	0044
10·0042	0041	0040	0039	0038	0037	0036	0035	0034	0033
10·0032	0032	0031	0030	0029	0028	0027	0026	0025	0025
10·0024	0023	0022	0022	0021	0020	0019	0019	0018	0017
10·0017	0016	0015	0015	0014	0013	0013	0012	0012	0011
10·0011	0010	0010	0009	0009	0008	0008	0007	0007	0006
10·0006	0006	0005	0005	0004	0004	0004	0004	0003	0003
10·0003	0002	0002	0002	0002	0001	0001	0001	0001	0001
10·0001	0001	0000	0000	0000	0000	0000	0000	0000	0000

	0'	6'	12'	18	24'	30'	36'	42'	48'	54'	1	2	3	4	5
0°	0000	0017	0035	0052	0070	0087	0105	0122	0140	0157	3	6	9	12	15
1	0175	0192	0209	0227	0244	0262	0279	0297	0314	0332	3	6	9	12	15
2	0349	0366	0384	0401	0419	0436	0454	0471	0488	0506	3	6	9	12	15
3	0523	0541	0558	0576	0593	0610	0628	0645	0663	0680	3	6	9	12	•15
4	0698	0715	0732	0750	0767	0785	0802	0819	0837	0854	3	6	9	12	15
5	0872	0889	0906	0924	0941	0958	0976	0993	1011	1028	3	6	9	12	14
6	1045	1063	1080	1097	1115	1132	1149	1167	1184	1201	3	6	9	12	14
7	1219	1236	1253	1271	1288	1305	1323	1340	1357	1374	3	6	9	12	14
8	1392	1409	1426	1444	1461	1478	1495	1513	1530	1547	3	6	9	12	14
9	1564	1582	1599	1616	1633	1650	1668	1685	1702	1719	3	6	9	12	14
10	1736	1754	1771	1788	1805	1822	1840	1857	1874	1891	3	6	9	12	14
11	1908	1925	1942	1959	1977	1994	2011	2028	2045	2062	3	6	9	11	14
12	2079	2096	2113	2130	2147	2164	2181	2198	2215	2232	3	6	9	11	14
13	2250	2267	2284	2300	2317	2334	2351	2368	2385	2402	3	6	8	11	14
14	2419	2436	2453	2470	2487	2504	2521	2538	2554	2571	3	6	8	11	14
15	2588	2605	2622	2639	2656	2672	2689	2706	2723	2740	3	6	8	11	14
16	2756	2773	2790	2807	2823	2840	2857	2874	2890	2907	3	6	8	11	14
17	2924	2940	2957	2974	2990	3007	3024	3040	3057	3074	3	6	8	11	14
18	3090	3107	3123	3140	3156	3173	3190	3206	3223	3239	3	6	8	11	14
19	3256	3272	3289	3305	3322	3338	3355	3371	3387	3404	3	5	8	11	14
20	3420	3437	3453	3469	3486	3502	3518	3535	3551	3567	3	5	8	11	14
21	3584	3600	3616	3633	3649	3665	3681	3697	3714	3730	3	5	8	11	14
22	3746	3762	3778	3795	3811	3827	3843	3859	3875	3891	3	5	8	11	14
23	3907	3923	3939	3955	3971	3987	4003	4019	4035	4051	3	5	8	11	14
24	4067	4083	4099	4115	4131	4147	4163	4179	4195	4210	3	5	8	11	13
25	4226	4242	4258	4274	4289	4305	4321	4337	4352	4368	3	5	8	11	13
26	4384	4399	4415	4431	4446	4462	4478	4493	4509	4524	3	5	8	10	13
27	4540	4555	4571	4586	4602	4617	4633	4648	4664	4679	3	5	8	10	13
28	4695	4710	4726	4741	4756	4772	4787	4802	4818	4833	3	5	8	10	13
29	4848	4863	4879	4894	4909	4924	4939	4955	4970	4985	3	5	8	10	13
30	5000	5015	5030	5045	5060	5075	5090	5105	5120	5135	3	5	8	10	13
31	5150	5165	5180	5195	5210	5225	5240	5255	5270	5284	2	5	7	10	12
32	5299	5314	5329	5344	5358	5373	5388	5402	5417	5432	2	5	7	10	12
33	5446	5461	5476	5490	5505	5519	5534	5548	5563	5577	2	5	7	10	12
34	5592	5606	5621	5635	5650	5664	5678	5693	5707	5721	2	5	7	10	12
35	5736	5750	5764	5779	5793	5807	5821	5835	5850	5864	2	5	7	10	12
36	5878	5892	5906	5920	5934	5948	5962	5976	5990	6004	2	5	7	9	12
37	6018	6032	6046	6060	6074	6088	6101	6115	6129	6143	2	5	7	9	12
38	6157	6170	6184	6198	6211	6225	6239	6252	6266	6280	2	5	7	9	11
39	6293	6307	6320	6334	6347	6361	6374	6388	6401	6414	2	4	7	9	11
40	6428	6441	6455	6468	6481	6494	6508	6521	6534	6547	2	4	7	9	11
41	6561	6574	6587	6600	6613	6626	6639	6652	6665	6678	2	4	7	9	11
42	6691	6704	6717	6730	6743	6756	6769	6782	6794	6807	2	4	6	9	11
43	6820	6833	6845	6858	6871	6884	6896	6909	6921	6934	2	4	6	8	11
44	6947	6959	6972	6984	6997	7009	7022	7034	7046	7059	2	4	6	8	10

NATURAL SINES.

	0′	6′	12′	18′	24′	30′	36′	42′	48′	54′
45°	7071	7083	7096	7108	7120	7133	7145	7157	7169	7181
46	7193	7206	7218	7230	7242	7254	7266	7278	7290	7302
47	7314	7325	7337	7349	7361	7373	7385	7396	7408	7420
48	7431	7443	7455	7466	7478	7490	7501	7513	7524	7536
49	7547	7558	7570	7581	7593	7604	7615	7627	7638	7649
50	7660	7672	7683	7694	7705	7716	7727	7738	7749	7760
51	7771	7782	7793	7804	7815	7826	7837	7848	7859	7869
52	7880	7891	7902	7912	7923	7934	7944	7955	7965	7976
53	7986	7997	8007	8018	8028	8039	8049	8059	8070	8080
54	8090	8100	8111	8121	8131	8141	8151	8161	8171	8181
55	8192	8202	8211	8221	8231	8241	8251	8261	8271	8281
56	8290	8300	8310	8320	8329	8339	8348	8358	8368	8377
57	8387	8396	8406	8415	8425	8434	8443	8453	8462	8471
58	8480	8490	8499	8508	8517	8526	8536	8545	8554	8563
59	8572	8581	8590	8599	8607	8616	8625	8634	8643	8652
60	8660	8669	8678	8686	8695	8704	8712	8721	8729	8738
61	8746	8755	8763	8771	8780	8788	8796	8805	8813	8821
62	8829	8838	8846	8854	8862	8870	8878	8886	8894	8902
63	8910	8918	8926	8934	8942	8949	8957	8965	8973	8980
64	8988	8996	9003	9011	9018	9026	9033	9041	9048	9056
65	9063	9070	9078	9085	9092	9100	9107	9114	9121	9128
66	9135	9143	9150	9157	9164	9171	9178	9184	9191	9198
67	9205	9212	9219	9225	9232	9239	9245	9252	9259	9265
68	9272	9278	9285	9291	9298	9304	9311	9317	9323	9330
69	9336	9342	9348	9354	9361	9367	9373	9379	9385	9391
70	9397	9403	9409	9415	9421	9426	9432	9438	9444	9449
71	9455	9461	9466	9472	9478	9483	9489	9494	9500	9505
72	9511	9516	9521	9527	9532	9537	9542	9548	9553	9558
73	9563	9568	9573	9578	9583	9588	9593	9598	9603	9608
74	9613	9617	9622	9627	9632	9636	9641	9646	9650	9655
75	9659	9664	9668	9673	9677	9681	9686	9690	9694	9699
76	9703	9707	9711	9715	9720	9724	9728	9732	9736	9740
77	9744	9748	9751	9755	9759	9763	9767	9770	9774	9778
78	9781	9785	9789	9792	9796	9799	9803	9806	9810	9813
79	9816	9820	9823	9826	9829	9833	9836	9839	9842	9845
80	9848	9851	9854	9857	9860	9863	9866	9869	9871	9874
81	9877	9880	9882	9885	9888	9890	9893	9895	9898	9900
82	9903	9905	9907	9910	9912	9914	9917	9919	9921	9923
83	9925	9928	9930	9932	9934	9936	9938	9940	9942	9943
84	9945	9947	9949	9951	9952	9954	9956	9957	9959	9960
85	9962	9963	9965	9966	9968	9969	9971	9972	9973	9974
86	9976	9977	9978	9979	9980	9981	9982	9983	9984	9985
87	9986	9987	9988	9989	9990	9990	9991	9992	9993	9993
88	9994	9995	9995	9996	9996	9997	9997	9997	9998	9998
89	9998	9999	9999	9999	9999	1·000 nearly.	1·000 nearly.	1·000 nearly.	1·000 nearly.	1·000 nearly.

	0'	6'	12'	18'	24'	30'	36'	42'	48'	54'	1	2	3	4	5
0°	1'000	1'000 nearly.	1'000 nearly.	1'000 nearly.	1'000 nearly.	9999	9999	9999	9999	9999	0	0	0	0	0
1	9998	9998	9998	9997	9997	9997	9996	9996	9995	9995	0	0	0	0	0
2	9994	9993	9993	9992	9991	9990	9990	9989	9988	9987	0	0	0	1	1
3	9986	9985	9984	9983	9982	9981	9980	9979	9978	9977	0	0	1	1	1
4	9976	9974	9973	9972	9971	9969	9968	9966	9965	9963	0	0	1	1	1
5	9962	9960	9959	9957	9956	9954	9952	9951	9949	9947	0	1	1	1	2
6	9945	9943	9942	9940	9938	9936	9934	9932	9930	.9928	0	1	1	1	2
7	9925	9923	9921	9919	9917	9914	9912	9910	9907	9905	0	1	1	2	2
8	9903	9900	9898	9895	9893	9890	9888	9885	9882	9880	0	1	1	2	2
9	9877	9874	9871	9869	9866	9863	9860	9857	9854	9851	0	1	1	2	2
10	9848	9845	9842	9839	9836	9833	9829	9826	9823	9820	1	1	2	2	3
11	9816	9813	9810	9806	9803	9799	9796	9792	9789	9785	1	1	2	2	3
12	9781	9778	9774	9770	9767	9763	9759	9755	9751	9748	1	1	2	3	3
13	9744	9740	9736	9732	9728	9724	9720	9715	9711	9707	1	1	2	3	3
14	9703	9699	9694	9690	9686	9681	9677	9673	9668	9664	1	1	2	3	4
15	9659	9655	9650	9646	9641	9636	9632	9627	9622	9617	1	2	2	3	4
16	9613	9608	9603	9598	9593	9588	9583	9578	9573	9568	1	2	2	3	4
17	9563	9558	9553	9548	9542	9537	9532	9527	9521	9516	1	2	3	4	4
18	9511	9505	9500	9494	9489	9483	9478	9472	9466	9461	1	2	3	4	5
19	9455	9449	9444	9438	9432	9426	9421	9415	9409	9403	1	2	3	4	5
20	9397	9391	9385	9379	9373	9367	9361	9354	9348	9342	1	2	3	4	5
21	9336	9330	9323	9317	9311	9304	9298	9291	9285	9278	1	2	3	4	5
22	9272	9265	9259	9252	9245	9239	9232	9225	9219	9212	1	2	3	4	6
23	9205	9198	9191	9184	9178	9171	9164	9157	9150	9143	1	2	3	5	6
24	9135	9128	9121	9114	9107	9100	9092	9085	9078	9070	1	2	4	5	6
25	9063	9056	9048	9041	9033	9026	9018	9011	9003	8996	1	3	4	5	6
26	8988	8980	8973	8965	8957	8949	8942	8934	8926	8918	1	3	4	5	6
27	8910	8902	8894	8886	8878	8870	8862	8854	8846	8838	1	3	4	5	7
28	8829	8821	8813	8805	8796	8788	8780	8771	8763	8755	1	3	4	6	7
29	8746	8738	8729	8721	8712	8704	8695	8686	8678	8669	1	3	4	6	7
30	8660	8652	8643	8634	8625	8616	8607	8599	8590	8581	1	3	4	6	7
31	8572	8563	8554	8545	8536	8526	8517	8508	8499	8490	2	3	5	6	8
32	8480	8471	8462	8453	8443	8434	8425	8415	8406	8396	2	3	5	6	8
33	8387	8377	8368	8358	8348	8339	8329	8320	8310	8300	2	3	5	6	8
34	8290	8281	8271	8261	8251	8241	8231	8221	8211	8202	2	3	5	7	8
35	8192	8181	8171	8161	8151	8141	8131	8121	8111	8100	2	3	5	7	8
36	8090	8080	8070	8059	8049	8039	8028	8018	8007	7997	2	3	5	7	9
37	7986	7976	7965	7955	7944	7934	7923	7912	7902	7891	2	4	5	7	9
38	7880	7869	7859	7848	7837	7826	7815	7804	7793	7782	2	4	5	7	9
39	7771	7760	7749	7738	7727	7716	7705	7694	7683	7672	2	4	6	7	9
40	7660	7649	7638	7627	7615	7604	7593	7581	7570	7559	2	4	6	8	9
41	7547	7536	7524	7513	7501	7490	7478	7466	7455	7443	2	4	6	8	10
42	7431	7420	7408	7396	7385	7373	7361	7349	7337	7325	2	4	6	8	10
43	7314	7302	7290	7278	7266	7254	7242	7230	7218	7206	2	4	6	8	10
44	7193	7181	7169	7157	7145	7133	7120	7108	7096	7083	2	4	6	8	10

N.B.—Numbers in difference columns to be subtracted, not added.—*See Rules.*

	0′	6′	12′	18′	24′	30′	36′	42′	48′	54′	1	2	3	4	5
45°	7071	7059	7046	7034	7022	7009	6997	6984	6972	6959	2	4	6	8	10
46	6947	6934	6921	6909	6896	6884	6871	6858	6845	6833	2	4	6	8	11
47	6820	6807	6794	6782	6769	6756	6743	6730	6717	6704	2	4	6	9	11
48	6691	6678	6665	6652	6639	6626	6613	6600	6587	6574	2	4	7	9	11
49	6561	6547	6534	6521	6508	6494	6481	6468	6455	6441	2	4	7	9	11
50	6428	6414	6401	6388	6374	6361	6347	6334	6320	6307	2	4	7	9	11
51	6293	6280	6266	6252	6239	6225	6211	6198	6184	6170	2	5	7	9	11
52	6157	6143	6129	6115	6101	6088	6074	6060	6046	6032	2	5	7	9	12
53	6018	6004	5990	5976	5962	5948	5934	5920	5906	5892	2	5	7	9	12
54	5878	5864	5850	5835	5821	5807	5793	5779	5764	5750	2	5	7	9	12
55	5736	5721	5707	5693	5678	5664	5650	5635	5621	5606	2	5	7	10	12
56	5592	5577	5563	5548	5534	5519	5505	5490	5476	5461	2	5	7	10	12
57	5446	5432	5417	5402	5388	5373	5358	5344	5329	5314	2	5	7	10	12
58	5299	5284	5270	5255	5240	5225	5210	5195	5180	5165	2	5	7	10	12
59	5150	5135	5120	5105	5090	5075	5060	5045	5030	5015	3	5	8	10	13
60	5000	4985	4970	4955	4939	4924	4909	4894	4879	4863	3	5	8	10	13
61	4848	4833	4818	4802	4787	4772	4756	4741	4726	4710	3	5	8	10	13
62	4695	4679	4664	4648	4633	4617	4602	4586	4571	4555	3	5	8	10	13
63	4540	4524	4509	4493	4478	4462	4446	4431	4415	4399	3	5	8	10	13
64	4384	4368	4352	4337	4321	4305	4289	4274	4258	4242	3	5	8	11	13
65	4226	4210	4195	4179	4163	4147	4131	4115	4099	4083	3	5	8	11	13
66	4067	4051	4035	4019	4003	3987	3971	3955	3939	3923	3	5	8	11	14
67	3907	3891	3875	3859	3843	3827	3811	3795	3778	3762	3	5	8	11	14
68	3746	3730	3714	3697	3681	3665	3649	3633	3616	3600	3	5	8	11	14
69	3584	3567	3551	3535	3518	3502	3486	3469	3453	3437	3	5	8	11	14
70	3420	3404	3387	3371	3355	3338	3322	3305	3289	3272	3	5	8	11	14
71	3256	3239	3223	3206	3190	3173	3156	3140	3123	3107	3	6	8	11	14
72	3090	3074	3057	3040	3024	3007	2990	2974	2957	2940	3	6	8	11	14
73	2924	2907	2890	2874	2857	2840	2823	2807	2790	2773	3	6	8	11	14
74	2756	2740	2723	2706	2689	2672	2656	2639	2622	2605	3	6	8	11	14
75	2588	2571	2554	2538	2521	2504	2487	2470	2453	2436	3	6	8	11	14
76	2419	2402	2385	2368	2351	2334	2317	2300	2284	2267	3	6	8	11	14
77	2250	2233	2215	2198	2181	2164	2147	2130	2113	2096	3	6	9	11	14
78	2079	2062	2045	2028	2011	1994	1977	1959	1942	1925	3	6	9	11	14
79	1908	1891	1874	1857	1840	1822	1805	1788	1771	1754	3	6	9	12	14
80	1736	1719	1702	1685	1668	1650	1633	1616	1599	1582	3	6	9	12	14
81	1564	1547	1530	1513	1495	1478	1461	1444	1426	1409	3	6	9	12	14
82	1392	1374	1357	1340	1323	1305	1288	1271	1253	1236	3	6	9	12	14
83	1219	1201	1184	1167	1149	1132	1115	1097	1080	1063	3	6	9	12	14
84	1045	1028	1011	0993	0976	0958	0941	0924	0906	0889	3	6	9	12	14
85	0872	0854	0837	0819	0802	0785	0767	0750	0732	0715	3	6	9	12	15
86	0698	0680	0663	0645	0628	0610	0593	0576	0558	0541	3	6	9	12	15
87	0523	0506	0488	0471	0454	0436	0419	0401	0384	0366	3	6	9	12	15
88	0349	0332	0314	0297	0279	0262	0244	0227	0209	0192	3	6	9	12	15
89	0175	0157	0140	0122	0105	0087	0070	0052	0035	0017	3	6	9	12	15

N.B.—Numbers in difference columns to be subtracted, not added.—*See Rules.*

	0′	6′	12′	18′	24′	30′	36′	42′	48′	54′	1	2	3	4	5
0°	·0000	0017	0035	0052	0070	0087	0105	0122	0140	0157	3	6	9	12	14
1	·0175	0192	0209	0227	0244	0262	0279	0297	0314	0332	3	6	9	12	15
2	·0349	0367	0384	0402	0419	0437	0454	0472	0489	0507	3	6	9	12	15
3	·0524	0542	0559	0577	0594	0612	0629	0647	0664	0682	3	6	9	12	15
4	·0699	0717	0734	0752	0769	0787	0805	0822	0840	0857	3	6	9	12	15
5	·0875	0892	0910	0928	0945	0963	0981	0998	1016	1033	3	6	9	12	15
6	·1051	1069	1086	1104	1122	1139	1157	1175	1192	1210	3	6	9	12	15
7	·1228	1246	1263	1281	1299	1317	1334	1352	1370	1388	3	6	9	12	15
8	·1405	1423	1441	1459	1477	1495	1512	1530	1548	1566	3	6	9	12	15
9	·1584	1602	1620	1638	1655	1673	1691	1709	1727	1745	3	6	9	12	15
10	·1763	1781	1799	1817	1835	1853	1871	1890	1908	1926	3	6	9	12	15
11	·1944	1962	1980	1998	2016	2035	2053	2071	2089	2107	3	6	9	12	15
12	·2126	2144	2162	2180	2199	2217	2235	2254	2272	2290	3	6	9	12	15
13	·2309	2327	2345	2364	2382	2401	2419	2438	2456	2475	3	6	9	12	15
14	·2493	2512	2530	2549	2568	2586	2605	2623	2642	2661	3	6	9	12	16
15	·2679	2698	2717	2736	2754	2773	2792	2811	2830	2849	3	6	9	13	16
16	·2867	2886	2905	2924	2943	2962	2981	3000	3019	3038	3	6	9	13	16
17	·3057	3076	3096	3115	3134	3153	3172	3191	3211	3230	3	6	10	13	16
18	·3249	3269	3288	3307	3327	3346	3365	3385	3404	3424	3	6	10	13	16
19	·3443	3463	3482	3502	3522	3541	3561	3581	3600	3620	3	6	10	13	17
20	·3640	3659	3679	3699	3719	3739	3759	3779	3799	3819	3	7	10	13	17
21	·3839	3859	3879	3899	3919	3939	3959	3979	4000	4020	3	7	10	13	17
22	·4040	4061	4081	4101	4122	4142	4163	4183	4204	4224	3	7	10	14	17
23	·4245	4265	4286	4307	4327	4348	4369	4390	4411	4431	3	7	10	14	17
24	·4452	4473	4494	4515	4536	4557	4578	4599	4621	4642	4	7	10	14	18
25	·4663	4684	4706	4727	4748	4770	4791	4813	4834	4856	4	7	11	14	18
26	·4877	4899	4921	4942	4964	4986	5008	5029	5051	5073	4	7	11	15	18
27	·5095	5117	5139	5161	5184	5206	5228	5250	5272	5295	4	7	11	15	18
28	·5317	5340	5362	5384	5407	5430	5452	5475	5498	5520	4	8	11	15	19
29	·5543	5566	5589	5612	5635	5658	5681	5704	5727	5750	4	8	12	15	19
30	·5774	5797	5820	5844	5867	5890	5914	5938	5961	5985	4	8	12	16	20
31	·6009	6032	6056	6080	6104	6128	6152	6176	6200	6224	4	8	12	16	20
32	·6249	6273	6297	6322	6346	6371	6395	6420	6445	6469	4	8	12	16	20
33	·6494	6519	6544	6569	6594	6619	6644	6669	6694	6720	4	8	13	17	21
34	·6745	6771	6796	6822	6847	6873	6899	6924	6950	6976	4	9	13	17	21
35	·7002	7028	7054	7080	7107	7133	7159	7186	7212	7239	4	9	13	18	22
36	·7265	7292	7319	7346	7373	7400	7427	7454	7481	7508	5	9	14	18	23
37	·7536	7563	7590	7618	7646	7673	7701	7729	7757	7785	5	9	14	18	23
38	·7813	7841	7869	7898	7926	7954	7983	8012	8040	8069	5	10	14	19	24
39	·8098	8127	8156	8185	8214	8243	8273	8302	8332	8361	5	10	15	20	24
40	·8391	8421	8451	8481	8511	8541	8571	8601	8632	8662	5	10	15	20	25
41	·8693	8724	8754	8785	8816	8847	8878	8910	8941	8972	5	10	16	21	26
42	·9004	9036	9067	9099	9131	9163	9195	9228	9260	9293	5	11	16	21	27
43	·9325	9358	9391	9424	9457	9490	9523	9556	9590	9623	6	11	17	22	28
44	·9657	9691	9725	9759	9793	9827	9861	9896	9930	9965	6	11	17	23	29

	0'	6'	12'	18'	24'	30'	36'	42'	48'	54'	1	2	3	4	5
45°	1·0000	0035	0070	0105	0141	0176	0212	0247	0283	0319	6	12	18	24	30
46	1·0355	0392	0428	0464	0501	0538	0575	0612	0649	0686	6	12	18	25	31
47	1·0724	0761	0799	0837	0875	0913	0951	0990	1028	1067	6	13	19	25	32
48	1·1106	1145	1184	1224	1263	1303	1343	1383	1423	1463	7	13	20	26	33
49	1·1504	1544	1585	1626	1667	1708	1750	1792	1833	1875	7	14	21	28	34
50	1·1918	1960	2002	2045	2088	2131	2174	2218	2261	2305	7	14	22	29	36
51	1·2349	2393	2437	2482	2527	2572	2617	2662	2708	2753	8	15	23	30	38
52	1·2799	2846	2892	2938	2985	3032	3079	3127	3175	3222	8	16	23	31	39
53	1·3270	3319	3367	3416	3465	3514	3564	3613	3663	3713	8	16	25	33	41
54	1·3764	3814	3865	3916	3968	4019	4071	4124	4176	4229	9	17	26	34	43
55	1·4281	4335	4388	4442	4496	4550	4605	4659	4715	4770	9	18	27	36	45
56	1·4826	4882	4938	4994	5051	5108	5166	5224	5282	5340	10	19	29	38	48
57	1·5399	5458	5517	5577	5637	5697	5757	5818	5880	5941	10	20	30	40	50
58	1·6003	6066	6128	6191	6255	6319	6383	6447	6512	6577	11	21	32	43	53
59	1·6643	6709	6775	6842	6909	6977	7045	7113	7182	7251	11	23	34	45	56
60	1·7321	7391	7461	7532	7603	7675	7747	7820	7893	7966	12	24	36	48	60
61	1·8040	8115	8190	8265	8341	8418	8495	8572	8650	8728	13	26	38	51	64
62	1·8807	8887	8967	9047	9128	9210	9292	9375	9458	9542	14	27	41	55	68
63	1·9626	9711	9797	9883	9970	0057	0145	0233	0323	0413	15	29	44	58	73
64	2·0503	0594	0686	0778	0872	0965	1060	1155	1251	1348	16	31	47	63	78
65	2·1445	1543	1642	1742	1842	1943	2045	2148	2251	2355	17	34	51	68	85
66	2·2460	2566	2673	2781	2889	2998	3109	3220	3332	3445	18	37	55	74	92
67	2·3559	3673	3789	3906	4023	4142	4262	4383	4504	4627	20	40	60	79	99
68	2·4751	4876	5002	5129	5257	5386	5517	5649	5782	5916	22	43	65	87	108
69	2·6051	6187	6325	6464	6605	6746	6889	7034	7179	7326	24	47	71	95	118
70	2·7475	7625	7776	7929	8083	8239	8397	8556	8716	8878	26	52	78	104	130
71	2·9042	9208	9375	9544	9714	9887	0061	0237	0415	0595	29	58	87	115	144
72	3·0777	0961	1146	1334	1524	1716	1910	2106	2305	2506	32	64	96	129	161
73	3·2709	2914	3122	3332	3544	3759	3977	4197	4420	4646	36	72	108	144	180
74	3·4874	5105	5339	5576	5816	6059	6305	6554	6806	7062	41	82	122	162	203
75	3·7321	7583	7848	8118	8391	8667	8947	9232	9520	9812	46	94	139	186	232
76	4·0108	0408	0713	1022	1335	1653	1976	2303	2635	2972	53	107	160	214	267
77	4·3315	3662	4015	4374	4737	5107	5483	5864	6252	6646	62	124	186	248	310
78	4·7046	7453	7867	8288	8716	9152	9594	0045	0504	0970	73	146	219	292	365
79	5·1446	1929	2422	2924	3435	3955	4486	5026	5578	6140	87	175	262	350	437
80	5·6713	7297	7894	8502	9124	9758	0405	1066	1742	2432					
81	6·3138	3859	4596	5350	6122	6912	7720	8548	9395	0264					
82	7·1154	2066	3002	3962	4947	5958	6996	8062	9158	0285					
83	8·1443	2636	3863	5126	6427	7769	9152	0579	2052	3572					
84	9·5144	9·677	9·845	10·02	10·20	10·39	10·58	10·78	10·99	11·20					
85	11·43	11·66	11·91	12·16	12·43	12·71	13·00	13·30	13·62	13·95					
86	14·30	14·67	15·06	15·46	15·89	16·35	16·83	17·34	17·89	18·46					
87	19·08	19·74	20·45	21·20	22·02	22·90	23·86	24·90	26·03	27·27					
88	28·64	30·14	31·82	33·69	35·80	38·19	40·92	44·07	47·74	52·08					
89	57·29	63·66	71·62	81·85	95·49	114·6	143·2	191·0	286·5	573·0					

Difference-columns cease to be useful, owing to the rapidity with which the value of the tangent changes.

NATURAL COTANGENTS.

	0′	6′	12′	18′	24′	30′	36′	42′	48′	54′	1	2	3	4	5
0°	Inf.	573·0	286·5	191·0	143·2	114·6	95·49	81·85	71·62	63·66					
1	57·29	52·08	47·74	44·07	40·92	38·19	35·80	33·69	31·82	30·14					
2	28·64	27·27	26·03	24·90	23·86	22·90	22·02	21·20	20·45	19·74					
3	19·08	18·46	17·89	17·34	16·83	16·35	15·89	15·46	15·06	14·67					
4	14·30	13·95	13·62	13·30	13·00	12·71	12·43	12·16	11·91	11·66					
5	11·43	11·20	10·99	10·78	10·58	10·39	10·20	10·02	9·845	9·677					
6	9·5144	3572	2052	0579	9152	7769	6427	5126	3863	2636					
7	8·1443	0285	9158	8062	6996	5958	4947	3962	3002	2066					
8	7·1154	0264	9395	8548	7720	6912	6122	5350	4596	3859					
9	6·3138	2432	1742	1066	0405	9758	9124	8502	7894	7297					
10	5·6713	6140	5578	5026	4486	3955	3435	2924	2422	1929					
11	5·1446	0970	0504	0045	9594	9152	8716	8288	7867	7453	74	148	222	296	370
12	4·7046	6646	6252	5864	5483	5107	4737	4374	4015	3662	63	125	188	252	314
13	4·3315	2972	2635	2303	1976	1653	1335	1022	0713	0408	53	107	160	214	267
14	4·0108	9812	9520	9232	8947	8667	8391	8118	7848	7583	46	93	139	186	232
15	3·7321	7062	6806	6554	6305	6059	5816	5576	5339	5105	41	82	122	163	204
16	3·4874	4646	4420	4197	3977	3759	3544	3332	3122	2914	36	72	108	144	180
17	3·2709	2506	2305	2106	1910	1716	1524	1334	1146	0961	32	64	96	129	161
18	3·0777	0595	0415	0237	0061	9887	9714	9544	9375	9208	29	58	87	115	144
19	2·9042	8878	8716	8556	8397	8239	8083	7929	7776	7625	26	52	78	104	130
20	2·7475	7326	7179	7034	6889	6746	6605	6464	6325	6187	24	47	71	95	118
21	2·6051	5916	5782	5649	5517	5386	5257	5129	5002	4876	22	43	65	87	108
22	2·4751	4627	4504	4383	4262	4142	4023	3906	3789	3673	20	40	60	79	99
23	2·3559	3445	3332	3220	3109	2998	2889	2781	2673	2566	18	37	55	74	92
24	2·2460	2355	2251	2148	2045	1943	1842	1742	1642	1543	17	34	51	68	85
25	2·1445	1348	1251	1155	1060	0965	0872	0778	0686	0594	16	31	47	63	78
26	2·0503	0413	0323	0233	0145	0057	9970	9883	9797	9711	15	29	44	58	73
27	1·9626	9542	9458	9375	9292	9210	9128	9047	8967	8887	14	27	41	55	68
28	1·8807	8728	8650	8572	8495	8418	8341	8265	8190	8115	13	26	38	51	64
29	1·8040	7966	7893	7820	7747	7675	7603	7532	7461	7391	12	24	36	48	60
30	1·7321	7251	7182	7113	7045	6977	6909	6842	6775	6709	11	23	34	45	56
31	1·6643	6577	6512	6447	6383	6319	6255	6191	6128	6066	11	21	32	43	53
32	1·6003	5941	5880	5818	5757	5697	5637	5577	5517	5458	10	20	30	40	50
33	1·5399	5340	5282	5224	5166	5108	5051	4994	4938	4882	10	19	29	38	48
34	1·4826	4770	4715	4659	4605	4550	4496	4442	4388	4335	9	18	27	36	45
35	1·4281	4229	4176	4124	4071	4019	3968	3916	3865	3814	9	17	26	34	43
36	1·3764	3713	3663	3613	3564	3514	3465	3416	3367	3319	8	16	25	33	41
37	1·3270	3222	3175	3127	3079	3032	2985	2938	2892	2846	8	16	23	31	39
38	1·2799	2753	2708	2662	2617	2572	2527	2482	2437	2393	8	15	23	30	38
39	1·2349	2305	2261	2218	2174	2131	2088	2045	2002	1960	7	14	22	29	36
40	1·1918	1875	1833	1792	1750	1708	1667	1626	1585	1544	7	14	21	28	34
41	1·1504	1463	1423	1383	1343	1303	1263	1224	1184	1145	7	13	20	26	33
42	1·1106	1067	1028	0990	0951	0913	0875	0837	0799	0761	6	13	19	25	32
43	1·0724	0686	0649	0612	0575	0538	0501	0464	0428	0392	6	12	18	25	31
44	1·0355	0319	0283	0247	0212	0176	0141	0105	0070	0035	6	12	18	24	30

Difference-columns not useful here, owing to the rapidity with which the value of the cotangent changes.

N.B.—Numbers in difference columns to be subtracted, not added.—See Rules.

	0′	6′	12′	18′	24′	30′	36′	42′	48′	54′	1	2	3	4	5
45°	1·0	0·9965	0·9930	0·9896	0·9861	0·9827	0·9793	0·9759	0·9725	0·9691	6	11	17	23	29
46	·9657	9623	9590	9556	9523	9490	9457	9424	9391	9358	6	11	17	22	28
47	·9325	9293	9260	9228	9195	9163	9131	9099	9067	9036	5	11	16	21	27
48	·9004	8972	8941	8910	8878	8847	8816	8785	8754	8724	5	10	16	21	26
49	·8693	8662	8632	8601	8571	8541	8511	8481	8451	8421	5	10	15	20	25
50	·8391	8361	8332	8302	8273	8243	8214	8185	8156	8127	5	10	15	20	24
51	·8098	8069	8040	8012	7983	7954	7926	7898	7869	7841	5	10	14	19	24
52	·7813	7785	7757	7729	7701	7673	7646	7618	7590	7563	5	9	14	18	23
53	·7536	7508	7481	7454	7427	7400	7373	7346	7319	7292	5	9	14	18	23
54	·7265	7239	7212	7186	7159	7133	7107	7080	7054	7028	4	9	13	18	22
55	·7002	6976	6950	6924	6899	6873	6847	6822	6796	6771	4	9	13	17	21
56	·6745	6720	6694	6669	6644	6619	6594	6569	6544	6519	4	8	13	17	21
57	·6494	6469	6445	6420	6395	6371	6346	6322	6297	6273	4	8	12	16	20
58	·6249	6224	6200	6176	6152	6128	6104	6080	6056	6032	4	8	12	16	20
59	·6009	5985	5961	5938	5914	5890	5867	5844	5820	5797	4	8	12	16	20
60	·5774	5750	5727	5704	5681	5658	5635	5612	5589	5566	4	8	12	15	19
61	·5543	5520	5498	5475	5452	5430	5407	5384	5362	5340	4	8	11	15	19
62	·5317	5295	5272	5250	5228	5206	5184	5161	5139	5117	4	7	11	15	18
63	·5095	5073	5051	5029	5008	4986	4964	4942	4921	4899	4	7	11	15	18
64	·4877	4856	4834	4813	4791	4770	4748	4727	4706	4684	4	7	11	14	18
65	·4663	4642	4621	4599	4578	4557	4536	4515	4494	4473	4	7	10	14	18
66	·4452	4431	4411	4390	4369	4348	4327	4307	4286	4265	3	7	10	14	17
67	·4245	4224	4204	4183	4163	4142	4122	4101	4081	4061	3	7	10	14	17
68	·4040	4020	4000	3979	3959	3939	3919	3899	3879	3859	3	7	10	13	17
69	·3839	3819	3799	3779	3759	3739	3719	3699	3679	3659	3	7	10	13	17
70	·3640	3620	3600	3581	3561	3541	3522	3502	3482	3463	3	6	10	13	17
71	·3443	3424	3404	3385	3365	3346	3327	3307	3288	3269	3	6	10	13	16
72	·3249	3230	3211	3191	3172	3153	3134	3115	3096	3076	3	6	10	13	16
73	·3057	3038	3019	3000	2981	2962	2943	2924	2905	2886	3	6	9	13	16
74	·2867	2849	2830	2811	2792	2773	2754	2736	2717	2698	3	6	9	13	16
75	·2679	2661	2642	2623	2605	2586	2568	2549	2530	2512	3	6	9	12	16
76	·2493	2475	2456	2438	2419	2401	2382	2364	2345	2327	3	6	9	12	15
77	·2309	2290	2272	2254	2235	2217	2199	2180	2162	2144	3	6	9	12	15
78	·2126	2107	2089	2071	2053	2035	2016	1998	1980	1962	3	6	9	12	15
79	·1944	1926	1908	1890	1871	1853	1835	1817	1799	1781	3	6	9	12	15
80	·1763	1745	1727	1709	1691	1673	1655	1638	1620	1602	3	6	9	12	15
81	·1584	1566	1548	1530	1512	1495	1477	1459	1441	1423	3	6	9	12	15
82	·1405	1388	1370	1352	1334	1317	1299	1281	1263	1246	3	6	9	12	15
83	·1228	1210	1192	1175	1157	1139	1122	1104	1086	1069	3	6	9	12	15
84	·1051	1033	1016	0998	0981	0963	0945	0928	0910	0892	3	6	9	12	15
85	·0875	0857	0840	0822	0805	0787	0769	0752	0734	0717	3	6	9	12	15
86	·0699	0682	0664	0647	0629	0612	0594	0577	0559	0542	3	6	9	12	15
87	·0524	0507	0489	0472	0454	0437	0419	0402	0384	0367	3	6	9	12	15
88	·0349	0332	0314	0297	0279	0262	0244	0227	0209	0192	3	6	9	12	15
89	·0175	0157	0140	0122	0105	0087	0070	0052	0035	0017	3	6	9	12	14

N.B.—Numbers in difference columns to be subtracted, not added.—*See Rules.*

	0′	6′	12′	18′	24′	30′	36′	42′	48′	54′	1	2	3	4	5
0°	1·0000	0000	0000	0000	0000	0000	0001	0001	0001	0001	0	0	0	0	0
1	1·0002	0002	0002	0003	0003	0003	0004	0004	0005	0006	0	0	0	0	0
2	1·0006	0007	0007	0008	0009	0010	0010	0011	0012	0013	0	0	0	0	0
3	1·0014	0015	0016	0017	0018	co19	0020	0021	0022	0023	0	0	1	1	1
4	1·0024	0026	0027	0028	0030	0031	0032	0034	0035	0037	0	0	1	1	1
5	1·0038	0040	0041	0043	0045	0046	0048	0050	0051	0053	0	1	1	1	1
6	1·0055	0057	0059	0061	0063	0065	0067	0069	0071	0073	0	1	1	1	2
7	1·0075	0077	0079	0082	0084	0086	0089	0091	0093	0096	0	1	1	2	2
8	1·0098	0101	0103	0106	0108	0111	0114	0116	0119	0122	0	1	1	2	2
9	1·0125	0127	0130	0133	0136	0139	0142	0145	0148	0151	0	1	1	2	2
10	1·0154	0157	0161	0164	0167	0170	0174	0177	0180	0184	1	1	2	2	3
11	1·0187	0191	0194	0198	0201	0205	0209	0212	0216	0220	1	1	2	3	3
12	1·0223	0227	0231	0235	0239	0243	0247	0251	0255	0259	1	1	2	3	3
13	1·0263	0267	0271	0276	0280	0284	0288	0293	0297	0302	1	1	2	3	4
14	1·0306	0311	0315	0320	0324	0329	0334	0338	0343	0348	1	2	2	3	4
15	1·0353	0358	0363	0367	0372	0377	0382	0388	0393	0398	1	2	3	3	4
16	1·0403	0408	0413	0419	0424	0429	0435	0440	0446	0451	1	2	3	4	5
17	1·0457	0463	0468	0474	0480	0485	0491	0497	0503	0509	1	2	3	4	5
18	1·0515	0521	0527	0533	0539	0545	0551	0557	0564	0570	1	2	3	4	5
19	1·0576	0583	0589	0595	0602	0608	0615	0622	0628	0635	1	2	3	4	5
20	1·0642	0649	0655	0662	0669	0676	0683	0690	0697	0704	1	2	3	5	6
21	1·0711	0719	0726	0733	0740	0748	0755	0763	0770	0778	1	2	4	5	6
22	1·0785	0793	0801	0808	0816	0824	0832	0840	0848	0856	1	3	4	5	6
23	1·0864	0872	0880	0888	0896	0904	0913	0921	0929	0938	1	3	4	6	7
24	1·0946	0955	0963	0972	0981	0989	0998	1007	1016	1025	1	3	4	6	7
25	1·1034	1043	1052	1061	1070	1079	1089	1098	1107	1117	2	3	5	6	8
26	1·1126	1136	1145	1155	1164	1174	1184	1194	1203	1213	2	3	5	6	8
27	1·1223	1233	1243	1253	1264	1274	1284	1294	1305	1315	2	3	5	7	9
28	1·1326	1336	1347	1357	1368	1379	1390	1401	1412	1423	2	4	5	7	9
29	1·1434	1445	1456	1467	1478	1490	1501	1512	1524	1535	2	4	6	8	9
30	1·1547	1559	1570	1582	1594	1606	1618	1630	1642	1654	2	4	6	8	10
31	1·1666	1679	1691	1703	1716	1728	1741	1753	1766	1779	2	4	6	8	10
32	1·1792	1805	1818	1831	1844	1857	1870	1883	1897	1910	2	4	7	9	11
33	1·1924	1937	1951	1964	1978	1992	2006	2020	2034	2048	2	5	7	9	12
34	1·2062	2076	2091	2105	2120	2134	2149	2163	2178	2193	2	5	7	10	12
35	1·2208	2223	2238	2253	2268	2283	2299	2314	2329	2345	3	5	8	10	13
36	1·2361	2376	2392	2408	2424	2440	2456	2472	2489	2505	3	5	8	11	13
37	1·2521	2538	2554	2571	2588	2605	2622	2639	2656	2673	3	6	8	11	14
38	1·2690	2708	2725	2742	2760	2778	2796	2813	2831	2849	3	6	9	12	15
39	1·2868	2886	2904	2923	2941	2960	2978	2997	3016	3035	3	6	9	12	16
40	1·3054	3073	3093	3112	3131	3151	3171	3190	3210	3230	3	7	10	13	16
41	1·3250	3270	3291	3311	3331	3352	3373	3393	3414	3435	3	7	10	14	17
42	1·3456	3478	3499	3520	3542	3563	3585	3607	3629	3651	4	7	11	14	18
43	1·3673	3696	3718	3741	3763	3786	3809	3832	3855	3878	4	8	11	15	19
44	1·3902	3925	3949	3972	3996	4020	4044	4069	4093	4118	4	8	12	16	20

	0′	6′	12′	18′	24′	30′	36′	42′	48′	54′	1	2	3	4	5
45°	1·4142	4167	4192	4217	4242	4267	4293	4318	4344	4370	4	8	13	17	21
46	1·4396	4422	4448	4474	4501	4527	4554	4581	4608	4635	4	9	13	18	22
47	1·4663	4690	4718	4746	4774	4802	4830	4859	4887	4916	5	9	14	19	23
48	1·4945	4974	5003	5032	5062	5092	5121	5151	5182	5212	5	10	15	20	25
49	1·5243	5273	5304	5335	5366	5398	5429	5461	5493	5525	5	10	16	21	26
50	1·5557	5590	5622	5655	5688	5721	5755	5788	5822	5856	6	11	17	22	28
51	1·5890	5925	5959	5994	6029	6064	6099	6135	6171	6207	6	12	18	24	29
52	1·6243	6279	6316	6353	6390	6427	6464	6502	6540	6578	6	12	19	25	31
53	1·6616	6655	6694	6733	6772	6812	6852	6892	6932	6972	7	13	20	26	33
54	1·7013	7054	7095	7137	7179	7221	7263	7305	7348	7391	7	14	21	28	35
55	1·7434	7478	7522	7566	7610	7655	7700	7745	7791	7837	7	15	22	30	37
56	1·7883	7929	7976	8023	8070	8118	8166	8214	8263	8312	8	16	24	32	40
57	1·8361	8410	8460	8510	8561	8612	8663	8714	8766	8818	9	17	26	34	43
58	1·8871	8924	8977	9031	9084	9139	9194	9249	9304	9360	9	18	27	36	45
59	1·9416	9473	9530	9587	9645	9703	9762	9821	9880	9940	10	19	29	39	49
60	2·0000	0061	0122	0183	0245	0308	0371	0434	0498	0562	10	21	31	42	52
61	2·0627	0692	0757	0824	0890	0957	1025	1093	1162	1231	11	22	34	45	56
62	2·1301	1371	1441	1513	1584	1657	1730	1803	1877	1952	12	24	36	48	61
63	2·2027	2103	2179	2256	2333	2412	2490	2570	2650	2730	13	26	39	52	66
64	2·2812	2894	2976	3060	3144	3228	3314	3400	3486	3574	14	28	43	57	71
65	2·3662	3751	3841	3931	4022	4114	4207	4300	4395	4490	15	31	46	62	77
66	2·4586	4683	4780	4879	4978	5078	5180	5282	5384	5488	17	34	50	67	84
67	2·5593	5699	5805	5913	6022	6131	6242	6354	6466	6580	18	37	55	73	92
68	2·6695	6811	6927	7046	7165	7285	7407	7529	7653	7778	20	40	60	81	101
69	2·7904	8032	8161	8291	8422	8555	8688	8824	8960	9099	22	44	67	89	111
70	2·9238	9379	9521	9665	9811	9957	‾0106	‾0256	‾0407	‾0561	25	49	74	99	123
71	3·0716	0872	1030	1190	1352	1515	1681	1848	2017	2188	27	55	82	110	137
72	3·2361	2535	2712	2891	3072	3255	3440	3628	3817	4009	31	61	92	123	154
73	3·4203	4399	4598	4799	5003	5209	5418	5629	5843	6060	35	69	104	138	173
74	3·6280	6502	6727	6955	7186	7420	7657	7897	8140	8387	39	79	118	157	196
75	3·8637	8890	9147	9408	9672	9939	‾0211	‾0486	‾0765	‾1048	45	90	135	180	225
76	4·1336	1627	1923	2223	2527	2837	3150	3469	3792	4121	52	104	156	208	260
77	4·4454	4793	5137	5486	5841	6202	6569	6942	7321	7706	61	122	182	243	304
78	4·8097	8496	8901	9313	9732	‾0159	‾0593	‾1034	‾1484	‾1942	72	144	216	287	359
79	5·2408	2883	3367	3860	4362	4874	5396	5928	6470	7023	86	173	259	345	432
80	5·7588	8164	8751	9351	9963	‾0589	‾1227	‾1880	‾2546	‾3228					
81	6·3925	4637	5366	6111	6874	7655	8454	9273	‾0112	‾0972					
82	7·1853	2757	3684	4635	5611	6613	7642	8700	9787	‾0905					
83	8·2055	3238	4457	5711	‾7004	8337	9711	‾1129	‾2593	‾4105					
84	9·5668	7283	8955	0685	2477	4334	6261	8260	11·03	11·25					
85	11·47	11·71	11·95	12·20	12·47	12·75	13·03	13·34	13·65	13·99					
86	14·34	14·70	15·09	15·50	15·93	16·38	16·86	17·37	17·91	18·49					
87	19·11	19·77	20·47	21·23	22·04	22·93	23·88	24·92	26·05	27·29					
88	28·65	30·16	31·84	33·71	35·81	38·20	40·93	44·08	47·75	52·09					
89	57·30	63·66	71·62	81·85	95·49	114·6	143·2	191·0	286·5	573·0					

NATURAL COSECANTS.

	0'	6'	12'	18'	24'	30'	36'	42'	48'	54'	1	2	3	4	5
0°	Inf.	573·0	286·5	191·0	143·2	114·6	95·49	81·85	71·62	63·66					
1	57·30	52·09	47·75	44·08	40·93	38·20	35·81	33·71	31·84	30·16					
2	28·65	27·29	26·05	24·92	23·88	22·93	22·04	21·23	20·47	19·77					
3	19·11	18·49	17·91	17·37	16·86	16·38	15·93	15·50	15·09	14·70					
4	14·34	13·99	13·65	13·34	13·03	12·75	12·47	12·20	11·95	11·71					
5	11·47	11·25	11·03	10·83	10·63	10·43	10·25	10·07	9·895	9·728					
6	9·5668	4105	2593	1129	9711	8337	7004	5711	4457	3238					
7	8·2055	0905	9787	8700	7642	6613	5611	4635	3684	2757					
8	7·1853	0972	0112	9273	8454	7655	6874	6111	5366	4637					
9	6·3925	3228	2546	1880	1227	0589	9963	9351	8751	8164					
10	5·7588	7023	6470	5928	5396	4874	4362	3860	3367	2883					
11	5·2408	1942	1484	1034	0593	0159	9732	9313	8901	8496					
12	4·8097	7706	7321	6942	6569	6202	5841	5486	5137	4793	61	122	182	243	304
13	4·4454	4121	3792	3469	3150	2837	2527	2223	1923	1627	52	104	156	208	260
14	4·1336	1048	0765	0486	0211	9939	9672	9408	9147	8890	45	90	135	180	225
15	3·8637	8387	8140	7897	7657	7420	7186	6955	6727	6502	39	79	118	157	196
16	3·6280	6060	5843	5629	5418	5209	5003	4799	4598	4399	35	69	104	138	173
17	3·4203	4009	3817	3628	3440	3255	3072	2891	2712	2535	31	61	92	123	154
18	3·2361	2188	2017	1848	1681	1515	1352	1190	1030	0872	27	55	82	110	137
19	3·0716	0561	0407	0256	0106	9957	9811	9665	9521	9379	25	49	74	99	123
20	2·9238	9099	8960	8824	8688	8555	8422	8291	8161	8032	22	44	67	89	111
21	2·7904	7778	7653	7529	7407	7285	7165	7046	6927	6811	20	40	60	81	101
22	2·6695	6580	6466	6354	6242	6131	6022	5913	5805	5699	18	37	55	73	92
23	2·5593	5488	5384	5282	5180	5078	4978	4879	4780	4683	17	34	50	67	84
24	2·4586	4490	4395	4300	4207	4114	4022	3931	3841	3751	15	31	46	62	77
25	2·3662	3574	3486	3400	3314	3228	3144	3060	2976	2894	14	28	43	57	71
26	2·2812	2730	2650	2570	2490	2412	2333	2256	2179	2103	13	26	39	52	65
27	2·2027	1952	1877	1803	1730	1657	1584	1513	1441	1371	12	24	36	48	60
28	2·1301	1231	1162	1093	1025	0957	0890	0824	0757	0692	11	22	34	45	56
29	2·0627	0562	0498	0434	0371	0308	0245	0183	0122	0061	10	21	31	42	52
30	2·0000	9940	9880	9821	9762	9703	9645	9587	9530	9473	10	19	29	39	49
31	1·9416	9360	9304	9249	9194	9139	9084	9031	8977	8924	9	18	27	36	45
32	1·8871	8818	8766	8714	8663	8612	8561	8510	8460	8410	8	17	25	34	42
33	1·8361	8312	8263	8214	8166	8118	8070	8023	7976	7929	8	16	24	32	40
34	1·7883	7837	7791	7745	7700	7655	7610	7566	7522	7478	7	15	22	30	37
35	1·7434	7391	7348	7305	7263	7221	7179	7137	7095	7054	7	14	21	28	35
36	1·7013	6972	6932	6892	6852	6812	6772	6733	6694	6655	7	13	20	26	33
37	1·6616	6578	6540	6502	6464	6427	6390	6353	6316	6279	6	12	19	25	31
38	1·6243	6207	6171	6135	6099	6064	6029	5994	5959	5925	6	12	18	23	29
39	1·5890	5856	5822	5788	5755	5721	5688	5655	5622	5590	6	11	17	22	28
40	1·5557	5525	5493	5461	5429	5398	5366	5335	5304	5273	5	10	16	21	26
41	1·5243	5212	5182	5151	5121	5092	5062	5032	5003	4974	5	10	15	20	25
42	1·4945	4916	4887	4859	4830	4802	4774	4746	4718	4690	5	9	14	19	23
43	1·4663	4635	4608	4581	4554	4527	4501	4474	4448	4422	4	9	13	18	22
44	1·4396	4370	4344	4318	4293	4267	4242	4217	4192	4167	4	8	13	17	21

N.B.—Numbers in difference columns to be subtracted, not added.—See Rules.

	0	6′	12′	18′	24′	30′	36′	42′	48′	54′	1	2	3	4	5
45°	1·4142	4118	4093	4069	4044	4020	3996	3972	3949	3925	4	8	12	16	20
46	1·3902	3878	3855	3832	3809	3786	3763	3741	3718	3696	4	8	11	15	19
47	1·3673	3651	3629	3607	3585	3563	3542	3520	3499	3478	4	7	11	14	18
48	1·3456	3435	3414	3393	3373	3352	3331	3311	3291	3270	3	7	10	14	17
49	1·3250	3230	3210	3190	3171	3151	3131	3112	3093	3073	3	7	10	13	16
50	1·3054	3035	3016	2997	2978	2960	2941	2923	2904	2886	3	6	9	12	15
51	1·2868	2849	2831	2813	2796	2778	2760	2742	2725	2708	3	6	9	12	15
52	1·2690	2673	2656	2639	2622	2605	2588	2571	2554	2538	3	6	8	11	14
53	1·2521	2505	2489	2472	2456	2440	2424	2408	2392	2376	3	5	8	11	13
54	1·2361	2345	2329	2314	2299	2283	2268	2253	2238	2223	3	5	8	10	13
55	1·2208	2193	2178	2163	2149	2134	2120	2105	2091	2076	2	5	7	10	12
56	1·2062	2048	2034	2020	2006	1992	1978	1964	1951	1937	2	5	7	9	12
57	1·1924	1910	1897	1883	1870	1857	1844	1831	1818	1805	2	4	7	9	11
58	1·1792	1779	1766	1753	1741	1728	1716	1703	1691	1679	2	4	6	8	10
59	1·1666	1654	1642	1630	1618	1606	1594	1582	1570	1559	2	4	6	8	10
60	1·1547	1535	1524	1512	1501	1490	1478	1467	1456	1445	2	4	6	8	9
61	1·1434	1423	1412	1401	1390	1379	1368	1357	1347	1336	2	4	5	7	9
62	1·1326	1315	1305	1294	1284	1274	1264	1253	1243	1233	2	3	5	7	9
63	1·1223	1213	1203	1194	1184	1174	1164	1155	1145	1136	2	3	5	6	8
64	1·1126	1117	1107	1098	1089	1079	1070	1061	1052	1043	2	3	5	6	8
65	1·1034	1025	1016	1007	0998	0989	0981	0972	0963	0955	1	3	4	6	7
66	1·0946	0938	0929	0921	0913	0904	0896	0888	0880	0872	1	3	4	6	7
67	1·0864	0856	0848	0840	0832	0824	0816	0808	0801	0793	1	3	4	5	7
68	1·0785	0778	0770	0763	0755	0748	0740	0733	0726	0719	1	2	4	5	6
69	1·0711	0704	0697	0690	0683	0676	0669	0662	0655	0649	1	2	3	5	6
70	1·0642	0635	0628	0622	0615	0608	0602	0595	0589	0583	1	2	3	4	5
71	1·0576	0570	0564	0557	0551	0545	0539	0533	0527	0521	1	2	3	4	5
72	1·0515	0509	0503	0497	0491	0485	0480	0474	0468	0463	1	2	3	4	5
73	1·0457	0451	0446	0440	0435	0429	0424	0419	0413	0408	1	2	3	4	4
74	1·0403	0398	0393	0388	0382	0377	0372	0367	0363	0358	1	2	2	3	4
75	1·0353	0348	0343	0338	0334	0329	0324	0320	0315	0311	1	2	2	3	4
76	1·0306	0302	0297	0293	0288	0284	0280	0276	0271	0267	1	1	2	3	4
77	1·0263	0259	0255	0251	0247	0243	0239	0235	0231	0227	1	1	2	3	3
78	1·0223	0220	0216	0212	0209	0205	0201	0198	0194	0191	1	1	2	3	3
79	1·0187	0184	0180	0177	0174	0170	0167	0164	0161	0157	1	1	2	2	3
80	1·0154	0151	0148	0145	0142	0139	0136	0133	0130	0127	0	1	1	2	2
81	1·0125	0122	0119	0116	0114	0111	0108	0106	0103	0101	0	1	1	2	2
82	1·0098	0096	0093	0091	0089	0086	0084	0082	0079	0077	0	1	1	2	2
83	1·0075	0073	0071	0069	0067	0065	0063	0061	0059	0057	0	1	1	1	2
84	1·0055	0053	0051	0050	0048	0046	0045	0043	0041	0040	0	1	1	1	1
85	1·0038	0037	0035	0034	0032	0031	0030	0028	0027	0026	0	0	1	1	1
86	1·0024	0023	0022	0021	0020	0019	0018	0017	0016	0015	0	0	0	1	1
87	1·0014	0013	0012	0011	0010	0010	0009	0008	0007	0007	0	0	0	1	1
88	1·0006	0006	0005	0004	0004	0003	0003	0003	0002	0002	0	0	0	0	0
89	1·0002	0001	0001	0001	0001	0000	0000	0000	0000	0000	0	0	0	0	0

N.B.—Numbers in difference columns to be subtracted, not added.—See Rules.

	0′	6′	12′	18′	24′	30′	36′	42′	48′	54′	1	2	3	4	5
0°	0·0000	0017	0035	0052	0070	0087	0105	0122	0140	0157	3	6	9	12	15
1	0·0175	0192	0209	0227	0244	0262	0279	0297	0314	0332	3	6	9	12	15
2	0·0349	0367	0384	0401	0419	0436	0454	0471	0489	0506	3	6	9	12	15
3	0·0524	0541	0559	0576	0593	0611	0628	0646	0663	0681	3	6	9	12	15
4	0·0698	0716	0733	0750	0768	0785	0803	0820	0838	0855	3	6	9	12	15
5	0·0873	0890	0908	0925	0942	0960	0977	0995	1012	1030	3	6	9	12	15
6	0·1047	1065	1082	1100	1117	1134	1152	1169	1187	1204	3	6	9	12	15
7	0·1222	1239	1257	1274	1292	1309	1326	1344	1361	1379	3	6	9	12	15
8	0·1396	1414	1431	1449	1466	1484	1501	1518	1536	1553	3	6	9	12	15
9	0·1571	1588	1606	1623	1641	1658	1676	1693	1710	1728	3	6	9	12	15
10	0·1745	1763	1780	1798	1815	1833	1850	1868	1885	1902	3	6	9	12	15
11	0·1920	1937	1955	1972	1990	2007	2025	2042	2059	2077	3	6	9	12	15
12	0·2094	2112	2129	2147	2164	2182	2199	2217	2234	2251	3	6	9	12	15
13	0·2269	2286	2304	2321	2339	2356	2374	2391	2409	2426	3	6	9	12	15
14	0·2443	2461	2478	2496	2513	2531	2548	2566	2583	2601	3	6	9	12	15
15	0·2618	2635	2653	2670	2688	2705	2723	2740	2758	2775	3	6	9	12	15
16	0·2793	2810	2827	2845	2862	2880	2897	2915	2932	2950	3	6	9	12	15
17	0·2967	2985	3002	3019	3037	3054	3072	3089	3107	3124	3	6	9	12	15
18	0·3142	3159	3176	3194	3211	3229	3246	3264	3281	3299	3	6	9	12	15
19	0·3316	3334	3351	3368	3386	3403	3421	3438	3456	3473	3	6	9	12	15
20	0·3491	3508	3526	3543	3560	3578	3595	3613	3630	3648	3	6	9	12	15
21	0·3665	3683	3700	3718	3735	3752	3770	3787	3805	3822	3	6	9	12	15
22	0·3840	3857	3875	3892	3910	3927	3944	3962	3979	3997	3	6	9	12	15
23	0·4014	4032	4049	4067	4084	4102	4119	4136	4154	4171	3	6	9	12	15
24	0·4189	4206	4224	4241	4259	4276	4294	4311	4328	4346	3	6	9	12	15
25	0·4363	4381	4398	4416	4433	4451	4468	4485	4503	4520	3	6	9	12	15
26	0·4538	4555	4573	4590	4608	4625	4643	4660	4677	4695	3	6	9	12	15
27	0·4712	4730	4747	4765	4782	4800	4817	4835	4852	4869	3	6	9	12	15
28	0·4887	4904	4922	4939	4957	4974	4992	5009	5027	5044	3	6	9	12	15
29	0·5061	5079	5096	5114	5131	5149	5166	5184	5201	5219	3	6	9	12	15
30	0·5236	5253	5271	5288	5306	5323	5341	5358	5376	5393	3	6	9	12	15
31	0·5411	5428	5445	5463	5480	5498	5515	5533	5550	5568	3	6	9	12	15
32	0·5585	5603	5620	5637	5655	5672	5690	5707	5725	5742	3	6	9	12	15
33	0·5760	5777	5794	5812	5829	5847	5864	5882	5899	5917	3	6	9	12	15
34	0·5934	5952	5969	5986	6004	6021	6039	6056	6074	6091	3	6	9	12	15
35	0·6109	6126	6144	6161	6178	6196	6213	6231	6248	6266	3	6	9	12	15
36	0·6283	6301	6318	6336	6353	6370	6388	6405	6423	6440	3	6	9	12	15
37	0·6458	6475	6493	6510	6528	6545	6562	6580	6597	6615	3	6	9	12	15
38	0·6632	6650	6667	6685	6702	6720	6737	6754	6772	6789	3	6	9	12	15
39	0·6807	6824	6842	6859	6877	6894	6912	6929	6946	6964	3	6	9	12	15
40	0·6981	6999	7016	7034	7051	7069	7086	7103	7121	7138	3	6	9	12	15
41	0·7156	7173	7191	7208	7226	7243	7261	7278	7295	7313	3	6	9	12	15
42	0·7330	7348	7365	7383	7400	7418	7435	7453	7470	7487	3	6	9	12	15
43	0·7505	7522	7540	7557	7575	7592	7610	7627	7645	7662	3	6	9	12	15
44	0·7679	7697	7714	7732	7749	7767	7784	7802	7819	7837	3	6	9	12	15

	0'	6'	12'	18'	24'	30'	36'	42'	48'	54'	1	2	3	4	5
45°	0·7854	7871	7889	7906	7924	7941	7959	7976	7994	8011	3	6	9	12	15
46	0·8029	8046	8063	8081	8098	8116	8133	8151	8168	8186	3	6	9	12	15
47	0·8203	8221	8238	8255	8273	8290	8308	8325	8343	8360	3	6	9	12	15
48	0·8378	8395	8412	8430	8447	8465	8482	8500	8517	8535	3	6	9	12	15
49	0·8552	8570	8587	8604	8622	8639	8657	8674	8692	8709	3	6	9	12	15
50	0·8727	8744	8762	8779	8796	8814	8831	8849	8866	8884	3	6	9	12	15
51	0·8901	8919	8936	8954	8971	8988	9006	9023	9041	9058	3	6	9	12	15
52	0·9076	9093	9111	9128	9146	9163	9180	9198	9215	9233	3	6	9	12	15
53	0·9250	9268	9285	9303	9320	9338	9355	9372	9390	9407	3	6	9	12	15
54	0·9425	9442	9460	9477	9495	9512	9529	9547	9564	9582	3	6	9	12	15
55	0·9599	9617	9634	9652	9669	9687	9704	9721	9739	9756	3	6	9	12	15
56	0·9774	9791	9809	9826	9844	9861	9879	9896	9913	9931	3	6	9	12	15
57	0·9948	9966	9983	0001	0018	0036	0053	0071	0088	0105	3	6	9	12	15
58	1·0123	0140	0158	0175	0193	0210	0228	0245	0263	0280	3	6	9	12	15
59	1·0297	0315	0332	0350	0367	0385	0402	0420	0437	0455	3	6	9	12	15
60	1·0472	0489	0507	0524	0542	0559	0577	0594	0612	0629	3	6	9	12	15
61	1·0647	0664	0681	0699	0716	0734	0751	0769	0786	0804	3	6	9	12	15
62	1·0821	0838	0856	0873	0891	0908	0926	0943	0961	0978	3	6	9	12	15
63	1·0996	1013	1030	1048	1065	1083	1100	1118	1135	1153	3	6	9	12	15
64	1·1170	1188	1205	1222	1240	1257	1275	1292	1310	1327	3	6	9	12	15
65	1·1345	1362	1380	1397	1414	1432	1449	1467	1484	1502	3	6	9	12	15
66	1·1519	1537	1554	1572	1589	1606	1624	1641	1659	1676	3	6	9	12	15
67	1·1694	1711	1729	1746	1764	1781	1798	1816	1833	1851	3	6	9	12	15
68	1·1868	1886	1903	1921	1938	1956	1973	1990	2008	2025	3	6	9	12	15
69	1·2043	2060	2078	2095	2113	2130	2147	2165	2182	2200	3	6	9	12	15
70	1·2217	2235	2252	2270	2287	2305	2322	2339	2357	2374	3	6	9	12	15
71	1·2392	2409	2427	2444	2462	2479	2497	2514	2531	2549	3	6	9	12	15
72	1·2566	2584	2601	2619	2636	2654	2671	2689	2706	2723	3	6	9	12	15
73	1·2741	2758	2776	2793	2811	2828	2846	2863	2881	2898	3	6	9	12	15
74	1·2915	2933	2950	2968	2985	3003	3020	3038	3055	3073	3	6	9	12	15
75	1·3090	3107	3125	3142	3160	3177	3195	3212	3230	3247	3	6	9	12	15
76	1·3265	3282	3299	3317	3334	3352	3369	3387	3404	3422	3	6	9	12	15
77	1·3439	3456	3474	3491	3509	3526	3544	3561	3579	3596	3	6	9	12	15
78	1·3614	3631	3648	3666	3683	3701	3718	3736	3753	3771	3	6	9	12	15
79	1·3788	3806	3823	3840	3858	3875	3893	3910	3928	3945	3	6	9	12	15
80	1·3963	3980	3998	4015	4032	4050	4067	4085	4102	4120	3	6	9	12	15
81	1·4137	4155	4172	4190	4207	4224	4242	4259	4277	4294	3	6	9	12	15
82	1·4312	4329	4347	4364	4382	4399	4416	4434	4451	4469	3	6	9	12	15
83	1·4486	4504	4521	4539	4556	4573	4591	4608	4626	4643	3	6	9	12	15
84	1·4661	4678	4696	4713	4731	4748	4765	4783	4800	4818	3	6	9	12	15
85	1·4835	4853	4870	4888	4905	4923	4940	4957	4975	4992	3	6	9	12	15
86	1·5010	5027	5045	5062	5080	5097	5115	5132	5149	5167	3	6	9	12	15
87	1·5184	5202	5219	5237	5254	5272	5289	5307	5324	5341	3	6	9	12	15
88	1·5359	5376	5394	5411	5429	5446	5464	5481	5499	5516	3	6	9	12	15
89	1·5533	5551	5568	5586	5603	5621	5638	5656	5673	5691	3	6	9	12	15

SQUARES.

6	7	8
1˙124	1˙145	1˙16
1˙346	1˙369	1˙39
1˙588	1˙613	1˙6:
1˙850	1˙877	1˙9(
2˙132	2˙161	2˙1(
2˙434	2˙465	2˙49
2˙756	2˙789	2˙8:

	0	1	2	3	4	5	6	7	8	9	1	2	3	4	5	6	7	8	9
5·5	30·25	30·36	30·47	30·58	30·69	30·80	30·91	31·02	31·14	31·25	1	2	3	4	6	7	8	9	10
5·6	31·36	31·47	31·58	31·70	31·81	31·92	32·04	32·15	32·26	32·38	1	2	3	5	6	7	8	9	10
5·7	32·49	32·60	32·72	32·83	32·95	33·06	33·18	33·29	33·41	33·52	1	2	3	5	6	7	8	9	10
5·8	33·64	33·76	33·87	33·99	34·11	34·22	34·34	34·46	34·57	34·69	1	2	4	5	6	7	8	9	11
5·9	34·81	34·93	35·05	35·16	35·28	35·40	35·52	35·64	35·76	35·88	1	2	4	5	6	7	8	10	11
6·0	36·00	36·12	36·24	36·36	36·48	36·60	36·72	36·84	36·97	37·09	1	2	4	5	6	7	9	10	11
6·1	37·21	37·33	37·45	37·58	37·70	37·82	37·95	38·07	38·19	38·32	1	2	4	5	6	7	9	10	11
6·2	38·44	38·56	38·69	38·81	38·94	39·06	39·19	39·31	39·44	39·56	1	3	4	5	6	8	9	10	11
6·3	39·69	39·82	39·94	40·07	40·20	40·32	40·45	40·58	40·70	40·83	1	3	4	5	6	8	9	10	11
6·4	40·96	41·09	41·22	41·34	41·47	41·60	41·73	41·86	41·99	42·12	1	3	4	5	6	8	9	10	12
6·5	42·25	42·38	42·51	42·64	42·77	42·90	43·03	43·16	43·30	43·43	1	3	4	5	7	8	9	10	12
6·6	43·56	43·69	43·82	43·96	44·09	44·22	44·36	44·49	44·62	44·76	1	3	4	5	7	8	9	11	12
6·7	44·89	45·02	45·16	45·29	45·43	45·56	45·70	45·83	45·97	46·10	1	3	4	5	7	8	9	11	12
6·8	46·24	46·38	46·51	46·65	46·79	46·92	47·06	47·20	47·33	47·47	1	3	4	5	7	8	10	11	12
6·9	47·61	47·75	47·89	48·02	48·16	48·30	48·44	48·58	48·72	48·86	1	3	4	6	7	8	10	11	13
7·0	49·00	49·14	49·28	49·42	49·56	49·70	49·84	49·98	50·13	50·27	1	3	4	6	7	8	10	11	13
7·1	50·41	50·55	50·69	50·84	50·98	51·12	51·27	51·41	51·55	51·70	1	3	4	6	7	9	10	11	13
7·2	51·84	51·98	52·13	52·27	52·42	52·56	52·71	52·85	53·00	53·14	1	3	4	6	7	9	10	12	13
7·3	53·29	53·44	53·58	53·73	53·88	54·02	54·17	54·32	54·46	54·61	1	3	4	6	7	9	10	12	13
7·4	54·76	54·91	55·06	55·20	55·35	55·50	55·65	55·80	55·95	56·10	1	3	4	6	7	9	10	12	13
7·5	56·25	56·40	56·55	56·70	56·85	57·00	57·15	57·30	57·46	57·61	2	3	5	6	8	9	11	12	14
7·6	57·76	57·91	58·06	58·22	58·37	58·52	58·68	58·83	58·98	59·14	2	3	5	6	8	9	11	12	14
7·7	59·29	59·44	59·60	59·75	59·91	60·06	60·22	60·37	60·53	60·68	2	3	5	6	8	9	11	12	14
7·8	60·84	61·00	61·15	61·31	61·47	61·62	61·78	61·94	62·09	62·25	2	3	5	6	8	9	11	13	14
7·9	62·41	62·57	62·73	62·88	63·04	63·20	63·36	63·52	63·68	63·84	2	3	5	6	8	10	11	13	14
8·0	64·00	64·16	64·32	64·48	64·64	64·80	64·96	65·12	65·29	65·45	2	3	5	6	8	10	11	13	14
8·1	65·61	65·77	65·93	66·10	66·26	66·42	66·59	66·75	66·91	67·08	2	3	5	7	8	10	11	13	15
8·2	67·24	67·40	67·57	67·73	67·90	68·06	68·23	68·39	68·56	68·72	2	3	5	7	8	10	12	13	15
8·3	68·89	69·06	69·22	69·39	69·56	69·72	69·89	70·06	70·22	70·39	2	3	5	7	8	10	12	13	15
8·4	70·56	70·73	70·90	71·06	71·23	71·40	71·57	71·74	71·91	72·08	2	3	5	7	8	10	12	14	15
8·5	72·25	72·42	72·59	72·76	72·93	73·10	73·27	73·44	73·62	73·79	2	3	5	7	9	10	12	14	15
8·6	73·96	74·13	74·30	74·48	74·65	74·82	75·00	75·17	75·34	75·52	2	3	5	7	9	10	12	14	16
8·7	75·69	75·86	76·04	76·21	76·39	76·56	76·74	76·91	77·09	77·26	2	4	5	7	9	11	12	14	16
8·8	77·44	77·62	77·79	77·97	78·15	78·32	78·50	78·68	78·85	79·03	2	4	5	7	9	11	12	14	16
8·9	79·21	79·39	79·57	79·74	79·92	80·10	80·28	80·46	80·64	80·82	2	4	5	7	9	11	13	14	16
9·0	81·00	81·18	81·36	81·54	81·72	81·90	82·08	82·26	82·45	82·63	2	4	5	7	9	11	13	14	16
9·1	82·81	82·99	83·17	83·36	83·54	83·72	83·91	84·09	84·27	84·46	2	4	5	7	9	11	13	15	16
9·2	84·64	84·82	85·01	85·19	85·38	85·56	85·75	85·93	86·12	86·30	2	4	6	7	9	11	13	15	17
9·3	86·49	86·68	86·86	87·05	87·24	87·42	87·61	87·80	87·98	88·17	2	4	6	7	9	11	13	15	17
9·4	88·36	88·55	88·74	88·92	89·11	89·30	89·49	89·68	89·87	90·06	2	4	6	8	9	11	13	15	17
9·5	90·25	90·44	90·63	90·82	91·01	91·20	91·39	91·58	91·78	91·97	2	4	6	8	10	11	13	15	17
9·6	92·16	92·35	92·54	92·74	92·93	93·12	93·32	93·51	93·70	93·90	2	4	6	8	10	12	14	15	17
9·7	94·09	94·28	94·48	94·67	94·87	95·06	95·26	95·45	95·65	95·84	2	4	6	8	10	12	14	16	18
9·8	96·04	96·24	96·43	96·63	96·83	97·02	97·22	97·42	97·61	97·81	2	4	6	8	10	12	14	16	18
9·9	98·01	98·21	98·41	98·60	98·80	99·00	99·20	99·40	99·60	99·80	2	4	6	8	10	12	14	16	18

	0	1	2	3	4	5	6	7	8	9	1	2	3	4	5	6	7	8	9
10	10·00	10·05	10·10	10·15	10·20	10·25	10·30	10·34	10·39	10·44	0	1	1	2	2	3	3	4	4
11	10·49	10·54	10·58	10·63	10·68	10·72	10·77	10·82	10·86	10·91	0	1	1	2	2	3	3	4	4
12	10·95	11·00	11·05	11·09	11·14	11·18	11·22	11·27	11·31	11·36	0	1	1	2	2	3	3	4	4
13	11·40	11·45	11·49	11·53	11·58	11·62	11·66	11·70	11·75	11·79	0	1	1	2	2	3	3	3	4
14	11·83	11·87	11·92	11·96	12·00	12·04	12·08	12·12	12·17	12·21	0	1	1	2	2	2	3	3	4
15	12·25	12·29	12·33	12·37	12·41	12·45	12·49	12·53	12·57	12·61	0	1	1	2	2	2	3	3	4
16	12·65	12·69	12·73	12·77	12·81	12·85	12·88	12·92	12·96	13·00	0	1	1	2	2	2	3	3	4
17	13·04	13·08	13·11	13·15	13·19	13·23	13·27	13·30	13·34	13·38	0	1	1	2	2	2	3	3	3
18	13·42	13·45	13·49	13·53	13·56	13·60	13·64	13·67	13·71	13·75	0	1	1	1	2	2	3	3	3
19	13·78	13·82	13·86	13·89	13·93	13·96	14·00	14·04	14·07	14·11	0	1	1	1	2	2	3	3	3
20	14·14	14·18	14·21	14·25	14·28	14·32	14·35	14·39	14·42	14·46	0	1	1	1	2	2	2	3	3
21	14·49	14·53	14·56	14·59	14·63	14·66	14·70	14·73	14·76	14·80	0	1	1	1	2	2	2	3	3
22	14·83	14·87	14·90	14·93	14·97	15·00	15·03	15·07	15·10	15·13	0	1	1	1	2	2	2	3	3
23	15·17	15·20	15·23	15·26	15·30	15·33	15·36	15·39	15·43	15·46	0	1	1	1	2	2	2	3	3
24	15·49	15·52	15·56	15·59	15·62	15·65	15·68	15·72	15·75	15·78	0	1	1	1	2	2	2	3	3
25	15·81	15·84	15·87	15·91	15·94	15·97	16·00	16·03	16·06	16·09	0	1	1	1	2	2	2	3	3
26	16·12	16·16	16·19	16·22	16·25	16·28	16·31	16·34	16·37	16·40	0	1	1	1	2	2	2	2	3
27	16·43	16·46	16·49	16·52	16·55	16·58	16·61	16·64	16·67	16·70	0	1	1	1	2	2	2	2	3
28	16·73	16·76	16·79	16·82	16·85	16·88	16·91	16·94	16·97	17·00	0	1	1	1	1	2	2	2	3
29	17·03	17·06	17·09	17·12	17·15	17·18	17·20	17·23	17·26	17·29	0	1	1	1	1	2	2	2	3
30	17·32	17·35	17·38	17·41	17·44	17·46	17·49	17·52	17·55	17·58	0	1	1	1	1	2	2	2	3
31	17·61	17·64	17·66	17·69	17·72	17·75	17·78	17·80	17·83	17·86	0	1	1	1	1	2	2	2	3
32	17·89	17·92	17·94	17·97	18·00	18·03	18·06	18·08	18·11	18·14	0	1	1	1	1	2	2	2	2
33	18·17	18·19	18·22	18·25	18·28	18·30	18·33	18·36	18·38	18·41	0	1	1	1	1	2	2	2	2
34	18·44	18·47	18·49	18·52	18·55	18·57	18·60	18·63	18·65	18·68	0	1	1	1	1	2	2	2	2
35	18·71	18·73	18·76	18·79	18·81	18·84	18·87	18·89	18·92	18·95	0	1	1	1	1	2	2	2	2
36	18·97	19·00	19·03	19·05	19·08	19·10	19·13	19·16	19·18	19·21	0	1	1	1	1	2	2	2	2
37	19·24	19·26	19·29	19·31	19·34	19·36	19·39	19·42	19·44	19·47	0	1	1	1	1	2	2	2	2
38	19·49	19·52	19·54	19·57	19·59	19·62	19·65	19·67	19·70	19·72	0	1	1	1	1	2	2	2	2
39	19·75	19·77	19·80	19·82	19·85	19·87	19·90	19·92	19·95	19·97	0	1	1	1	1	2	2	2	2
40	20·00	20·02	20·05	20·07	20·10	20·12	20·15	20·17	20·20	20·22	0	0	1	1	1	1	2	2	2
41	20·25	20·27	20·30	20·32	20·35	20·37	20·40	20·42	20·45	20·47	0	0	1	1	1	1	2	2	2
42	20·49	20·52	20·54	20·57	20·59	20·62	20·64	20·66	20·69	20·71	0	0	1	1	1	1	2	2	2
43	20·74	20·76	20·78	20·81	20·83	20·86	20·88	20·90	20·93	20·95	0	0	1	1	1	1	2	2	2
44	20·98	21·00	21·02	21·05	21·07	21·10	21·12	21·14	21·17	21·19	0	0	1	1	1	1	2	2	2
45	21·21	21·24	21·26	21·28	21·31	21·33	21·35	21·38	21·40	21·42	0	0	1	1	1	1	2	2	2
46	21·45	21·47	21·49	21·52	21·54	21·56	21·59	21·61	21·63	21·66	0	0	1	1	1	1	2	2	2
47	21·68	21·70	21·73	21·75	21·77	21·79	21·82	21·84	21·86	21·89	0	0	1	1	1	1	2	2	2
48	21·91	21·93	21·95	21·98	22·00	22·02	22·05	22·07	22·09	22·11	0	0	1	1	1	1	2	2	2
49	22·14	22·16	22·18	22·20	22·23	22·25	22·27	22·29	22·32	22·34	0	0	1	1	1	1	2	2	2
50	22·36	22·38	22·41	22·43	22·45	22·47	22·49	22·52	22·54	22·56	0	0	1	1	1	1	2	2	2
51	22·58	22·61	22·63	22·65	22·67	22·69	22·72	22·74	22·76	22·78	0	0	1	1	1	1	2	2	2
52	22·80	22·83	22·85	22·87	22·89	22·91	22·93	22·96	22·98	23·00	0	0	1	1	1	1	2	2	2
53	23·02	23·04	23·07	23·09	23·11	23·13	23·15	23·17	23·19	23·22	0	0	1	1	1	1	2	2	2
54	23·24	23·26	23·28	23·30	23·32	23·35	23·37	23·39	23·41	23·43	0	0	1	1	1	1	1	2	2

	0	1	2	3	4	5	6	7	8	9	1	2	3	4	5	6	7	8	9
55	23·45	23·47	23·49	23·52	23·54	23·56	23·58	23·60	23·62	23·64	0	0	1	1	1	1	1	2	2
56	23·66	23·69	23·71	23·73	23·75	23·77	23·79	23·81	23·83	23·85	0	0	1	1	1	1	1	2	2
57	23·87	23·90	23·92	23·94	23·96	23·98	24·00	24·02	24·04	24·06	0	0	1	1	1	1	1	2	2
58	24·08	24·10	24·12	24·15	24·17	24·19	24·21	24·23	24·25	24·27	0	0	1	1	1	1	1	2	2
59	24·29	24·31	24·33	24·35	24·37	24·39	24·41	24·43	24·45	24·47	0	0	1	1	1	1	1	2	2
60	24·49	24·52	24·54	24·56	24·58	24·60	24·62	24·64	24·66	24·68	0	0	1	1	1	1	1	2	2
61	24·70	24·72	24·74	24·76	24·78	24·80	24·82	24·84	24·86	24·88	0	0	1	1	1	1	1	2	2
62	24·90	24·92	24·94	24·96	24·98	25·00	25·02	25·04	25·06	25·08	0	0	1	1	1	1	1	2	2
63	25·10	25·12	25·14	25·16	25·18	25·20	25·22	25·24	25·26	25·28	0	0	1	1	1	1	1	2	2
64	25·30	25·32	25·34	25·36	25·38	25·40	25·42	25·44	25·46	25.48	0	0	1	1	1	1	1	2	2
65	25·50	25·51	25·53	25·55	25·57	25·59	25·61	25·63	25·65	25·67	0	0	1	1	1	1	1	2	2
66	25·69	25·71	25·73	25·75	25·77	25·79	25·81	25·83	25·85	25·87	0	0	1	1	1	1	1	2	2
67	25·88	25·90	25·92	25·94	25·96	25·98	26·00	26·02	26·04	26·06	0	0	1	1	1	1	1	2	2
68	26·08	26·10	26·12	26·13	26·15	26·17	26·19	26·21	26·23	26·25	0	0	1	1	1	1	1	2	2
69	26·27	26·29	26·31	26·32	26·34	26·36	26·38	26·40	26·42	26·44	0	0	1	1	1	1	1	2	2
70	26·46	26·48	26·50	26·51	26·53	26·55	26·57	26·59	26·61	26·63	0	0	1	1	1	1	1	2	2
71	26·65	26·66	26·68	26·70	26·72	26·74	26·76	26·78	26·80	26·81	0	0	1	1	1	1	1	1	2
72	26·83	26·85	26·87	26·89	26·91	26·93	26·94	26·96	26·98	27·00	0	0	1	1	1	1	1	1	2
73	27·02	27·04	27·06	27·07	27·09	27·11	27·13	27·15	27·17	27·18	0	0	1	1	1	1	1	1	2
74	27·20	27·22	27·24	27·26	27·28	27·29	27·31	27·33	27·35	27·37	0	0	1	1	1	1	1	1	2
75	27·39	27·40	27·42	27·44	27·46	27·48	27·50	27·51	27·53	27·55	0	0	1	1	1	1	1	1	2
76	27·57	27·59	27·60	27·62	27·64	27·66	27·68	27·69	27·71	27·73	0	0	1	1	1	1	1	1	2
77	27·75	27·77	27·78	27·80	27·82	27·84	27·86	27·87	27·89	27·91	0	0	1	1	1	1	1	1	2
78	27·93	27·95	27·96	27·98	28·00	28·02	28·04	28·05	28·07	28·09	0	0	1	1	1	1	1	1	2
79	28·11	28·12	28·14	28·16	28·18	28·20	28·21	28·23	28·25	28·27	0	0	1	1	1	1	1	1	2
80	28·28	28·30	28·32	28·34	28·35	28·37	28·39	28·41	28·43	28·44	0	0	1	1	1	1	1	1	2
81	28·46	28·48	28·50	28·51	28·53	28·55	28·57	28·58	28·60	28·62	0	0	1	1	1	1	1	1	2
82	28·64	28·65	28·67	28·69	28·71	28·72	28·74	28·76	28·77	28·79	0	0	1	1	1	1	1	1	2
83	28·81	28·83	28·84	28·86	28·88	28·90	28·91	28·93	28·95	28·97	0	0	1	1	1	1	1	1	2
84	28·98	29·00	29·02	29·03	29·05	29·07	29·09	29·10	29·12	29·14	0	0	1	1	1	1	1	1	2
85	29·15	29·17	29·19	29·21	29·22	29·24	29·26	29·27	29·29	29·31	0	0	1	1	1	1	1	1	2
86	29·33	29·34	29·36	29·38	29·39	29·41	29·43	29·44	29·46	29·48	0	0	1	1	1	1	1	1	2
87	29·50	29·51	29·53	29·55	29·56	29·58	29·60	29·61	29·63	29·65	0	0	1	1	1	1	1	1	2
88	29·66	29·68	29·70	29·72	29·73	29·75	29·77	29·78	29·80	29·82	0	0	1	1	1	1	1	1	2
89	29·83	29·85	29·87	29·88	29·90	29·92	29·93	29·95	29·97	29·98	0	0	1	1	1	1	1	1	2
90	30·00	30·02	30·03	30·05	30·07	30·08	30·10	30·12	30·13	30·15	0	0	0	1	1	1	1	1	1
91	30·17	30·18	30·20	30·22	30·23	30·25	30·27	30·28	30·30	30·32	0	0	0	1	1	1	1	1	1
92	30·33	30·35	30·36	30·38	30·40	30·41	30·43	30·45	30·46	30·48	0	0	0	1	1	1	1	1	1
93	30·50	30·51	30·53	30·55	30·56	30·58	30·59	30·61	30·63	30·64	0	0	0	1	1	1	1	1	1
94	30·66	30·68	30·69	30·71	30·72	30·74	30·76	30·77	30·79	30·81	0	0	0	1	1	1	1	1	1
95	30·82	30·84	30·85	30·87	30·89	30·90	30·92	30·94	30·95	30·97	0	0	0	1	1	1	1	1	1
96	30·98	31·00	31·02	31·03	31·05	31·06	31·08	31·10	31·11	31·13	0	0	0	1	1	1	1	1	1
97	31·14	31·16	31·18	31·19	31·21	31·22	31·24	31·26	31·27	31·29	0	0	0	1	1	1	1	1	1
98	31·30	31·32	31·34	31·35	31·37	31·38	31·40	31·42	31·43	31·45	0	0	0	1	1	1	1	1	1
99	31·46	31·48	31·50	31·51	31·53	31·54	31·56	31·58	31·59	31·61	0	0	0	1	1	1	1	1	1

D

	0	1	2	3	4	5	6	7	8	9	1 2 3	4 5 6	7 8 9
10	31·62	31·78	31·94	32·09	32·25	32·40	32·56	32·71	32·86	33·02	2 3 5	6 8 9	11 12 14
11	33·17	33·32	33·47	33·62	33·76	33·91	34·06	34·21	34·35	34·50	1 3 4	6 7 9	10 12 13
12	34·64	34·79	34·93	35·07	35·21	35·36	35·50	35·64	35·78	35·92	1 3 4	6 7 8	10 11 13
13	36·06	36·19	36·33	36·47	36·61	36·74	36·88	37·01	37·15	37·28	1 3 4	5 7 8	10 11 12
14	37·42	37·55	37·68	37·82	37·95	38·08	38·21	38·34	38·47	38·60	1 3 4	5 7 8	9 11 12
15	38·73	38·86	38·99	39·12	39·24	39·37	39·50	39·62	39·75	39·87	1 3 4	5 6 8	9 10 11
16	40·00	40·12	40·25	40·37	40·50	40·62	40·74	40·87	40·99	41·11	1 2 4	5 6 7	9 10 11
17	41·23	41·35	41·47	41·59	41·71	41·83	41·95	42·07	42·19	42·31	1 2 4	5 6 7	8 10 11
18	42·43	42·54	42·66	42·78	42·90	43·01	43·13	43·24	43·36	43·47	1 2 3	5 6 7	8 9 10
19	43·59	43·70	43·82	43·93	44·05	44·16	44·27	44·38	44·50	44·61	1 2 3	5 6 7	8 9 10
20	44·72	44·83	44·94	45·06	45·17	45·28	45·39	45·50	45·61	45·72	1 2 3	4 6 7	8 9 10
21	45·83	45·93	46·04	46·15	46·26	46·37	46·48	46·58	46·69	46·80	1 2 3	4 5 6	8 9 10
22	46·90	47·01	47·12	47·22	47·33	47·43	47·54	47·64	47·75	47·85	1 2 3	4 5 6	7 8 9
23	47·96	48·06	48·17	48·27	48·37	48·48	48·58	48·68	48·79	48·89	1 2 3	4 5 6	7 8 9
24	48·99	49·09	49·19	49·30	49·40	49·50	49·60	49·70	49·80	49·90	1 2 3	4 5 6	7 8 9
25	50·00	50·10	50·20	50·30	50·40	50·50	50·60	50·70	50·79	50·89	1 2 3	4 5 6	7 8 9
26	50·99	51·09	51·19	51·28	51·38	51·48	51·58	51·67	51·77	51·87	1 2 3	4 5 6	7 8 9
27	51·96	52·06	52·15	52·25	52·35	52·44	52·54	52·63	52·73	52·82	1 2 3	4 5 6	7 8 9
28	52·92	53·01	53·10	53·20	53·29	53·39	53·48	53·57	53·67	53·76	1 2 3	4 5 6	7 7 8
29	53·85	53·94	54·04	54·13	54·22	54·31	54·41	54·50	54·59	54·68	1 2 3	4 5 5	6 7 8
30	54·77	54·86	54·95	55·05	55·14	55·23	55·32	55·41	55·50	55·59	1 2 3	4 4 5	6 7 8
31	55·68	55·77	55·86	55·95	56·04	56·12	56·21	56·30	56·39	56·48	1 2 3	3 4 5	6 7 8
32	56·57	56·66	56·75	56·83	56·92	57·01	57·10	57·18	57·27	57·36	1 2 3	3 4 5	6 7 8
33	57·45	57·53	57·62	57·71	57·79	57·88	57·97	58·05	58·14	58·22	1 2 3	3 4 5	6 7 8
34	58·31	58·40	58·48	58·57	58·65	58·74	58·82	58·91	58·99	59·08	1 2 3	3 4 5	6 7 8
35	59·16	59·25	59·33	59·41	59·50	59·58	59·67	59·75	59·83	59·92	1 2 2	3 4 5	6 7 8
36	60·00	60·08	60·17	60·25	60·33	60·42	60·50	60·58	60·66	60·75	1 2 2	3 4 5	6 7 7
37	60·83	60·91	60·99	61·07	61·16	61·24	61·32	61·40	61·48	61·56	1 2 2	3 4 5	6 7 7
38	61·64	61·73	61·81	61·89	61·97	62·05	62·13	62·21	62·29	62·37	1 2 2	3 4 5	6 6 7
39	62·45	62·53	62·61	62·69	62·77	62·85	62·93	63·01	63·09	63·17	1 2 2	3 4 5	6 6 7
40	63·25	63·32	63·40	63·48	63·56	63·64	63·72	63·80	63·87	63·95	1 2 2	3 4 5	6 6 7
41	64·03	64·11	64·19	64·27	64·34	64·42	64·50	64·58	64·65	64·73	1 2 2	3 4 5	5 6 7
42	64·81	64·88	64·96	65·04	65·12	65·19	65·27	65·35	65·42	65·50	1 2 2	3 4 5	5 6 7
43	65·57	65·65	65·73	65·80	65·88	65·95	66·03	66·11	66·18	66·26	1 2 2	3 4 5	5 6 7
44	66·33	66·41	66·48	66·56	66·63	66·71	66·78	66·86	66·93	67·01	1 2 2	3 4 5	5 6 7
45	67·08	67·16	67·23	67·31	67·38	67·45	67·53	67·60	67·68	67·75	1 1 2	3 4 4	5 6 7
46	67·82	67·90	67·97	68·04	68·12	68·19	68·26	68·34	68·41	68·48	1 1 2	3 4 4	5 6 7
47	68·56	68·63	68·70	68·77	68·85	68·92	68·99	69·07	69·14	69·21	1 1 2	3 4 4	5 6 7
48	69·28	69·35	69·43	69·50	69·57	69·64	69·71	69·79	69·86	69·93	1 1 2	3 4 4	5 6 6
49	70·00	70·07	70·14	70·21	70·29	70·36	70·43	70·50	70·57	70·64	1 1 2	3 4 4	5 6 6
50	70·71	70·78	70·85	70·92	70·99	71·06	71·13	71·20	71·27	71·34	1 1 2	3 4 4	5 6 6
51	71·41	71·48	71·55	71·62	71·69	71·76	71·83	71·90	71·97	72·04	1 1 2	3 4 4	5 6 6
52	72·11	72·18	72·25	72·32	72·39	72·46	72·53	72·59	72·66	72·73	1 1 2	3 3 4	5 6 6
53	72·80	72·87	72·94	73·01	73·08	73·14	73·21	73·28	73·35	73·42	1 1 2	3 3 4	5 5 6
54	73·48	73·55	73·62	73·69	73·76	73·82	73·89	73·96	74·03	74·09	1 1 2	3 3 4	5 5 6

	0	1	2	3	4	5	6	7	8	9	1 2 3	4 5 6	7 8 9
55	74·16	74·23	74·30	74·36	74·43	74·50	74·57	74·63	74·70	74·77	1 1 2	3 3 4	5 5 6
56	74·83	74·90	74·97	75·03	75·10	75·17	75·23	75·30	75·37	75·43	1 1 2	3 3 4	5 5 6
57	75·50	75·56	75·63	75·70	75·76	75·83	75·89	75·96	76·03	76·09	1 1 2	3 3 4	5 5 6
58	76·16	76·22	76·29	76·35	76·42	76·49	76·55	76·62	76·68	76·75	1 1 2	3 3 4	5 5 6
59	76·81	76·88	76·94	77·01	77·07	77·14	77·20	77·27	77·33	77·40	1 1 2	3 3 4	4 5 6
60	77·46	77·52	77·59	77·65	77·72	77·78	77·85	77·91	77·97	78·04	1 1 2	3 3 4	4 5 6
61	78·10	78·17	78·23	78·29	78·36	78·42	78·49	78·55	78·61	78·68	1 1 2	3 3 4	4 5 6
62	78·74	78·80	78·87	78·93	78·99	79·06	79·12	79·18	79·25	79·31	1 1 2	3 3 4	4 5 6
63	79·37	79·44	79·50	79·56	79·62	79·69	79·75	79·81	79·87	79·94	1 1 2	3 3 4	4 5 6
64	80·00	80·06	80·12	80·19	80·25	80·31	80·37	80·44	80·50	80·56	1 1 2	2 3 4	4 5 6
65	80·62	80·68	80·75	80·81	80·87	80·93	80·99	81·06	81·12	81·18	1 1 2	2 3 4	4 5 5
66	81·24	81·30	81·36	81·42	81·49	81·55	81·61	81·67	81·73	81·79	1 1 2	2 3 4	4 5 5
67	81·85	81·91	81·98	82·04	82·10	82·16	82·22	82·28	82·34	82·40	1 1 2	2 3 4	4 5 5
68	82·46	82·52	82·58	82·64	82·70	82·76	82·83	82·89	82·95	83·01	1 1 2	2 3 4	4 5 5
69	83·07	83·13	83·19	83·25	83·31	83·37	83·43	83·49	83·55	83·61	1 1 2	2 3 4	4 5 5
70	83·67	83·73	83·79	83·85	83·90	83·96	84·02	84·08	84·14	84·20	1 1 2	2 3 4	4 5 5
71	84·26	84·32	84·38	84·44	84·50	84·56	84·62	84·68	84·73	84·79	1 1 2	2 3 4	4 5 5
72	84·85	84·91	84·97	85·03	85·09	85·15	85·21	85·26	85·32	85·38	1 1 2	2 3 3	4 5 5
73	85·44	85·50	85·56	85·62	85·67	85·73	85·79	85·85	85·91	85·97	1 1 2	2 3 3	4 5 5
74	86·02	86·08	86·14	86·20	86·26	86·31	86·37	86·43	86·49	86·54	1 1 2	2 3 3	4 5 5
75	86·60	86·66	86·72	86·78	86·83	86·89	86·95	87·01	87·06	87·12	1 1 2	2 3 3	4 5 5
76	87·18	87·24	87·29	87·35	87·41	87·46	87·52	87·58	87·64	87·69	1 1 2	2 3 3	4 5 5
77	87·75	87·81	87·86	87·92	87·98	88·03	88·09	88·15	88·20	88·26	1 1 2	2 3 3	4 4 5
78	88·32	88·37	88·43	88·49	88·54	88·60	88·66	88·71	88·77	88·83	1 1 2	2 3 3	4 4 5
79	88·88	88·94	88·99	89·05	89·11	89·16	89·22	89·27	89·33	89·39	1 1 2	2 3 3	4 4 5
80	89·44	89·50	89·55	89·61	89·67	89·72	89·78	89·83	89·89	89·94	1 1 2	2 3 3	4 4 5
81	90·00	90·06	90·11	90·17	90·22	90·28	90·33	90·39	90·44	90·50	1 1 2	2 3 3	4 4 5
82	90·55	90·61	90·66	90·72	90·77	90·83	90·88	90·94	90·99	91·05	1 1 2	2 3 3	4 4 5
83	91·10	91·16	91·21	91·27	91·32	91·38	91·43	91·49	91·54	91·60	1 1 2	2 3 3	4 4 5
84	91·65	91·71	91·76	91·82	91·87	91·92	91·98	92·03	92·09	92·14	1 1 2	2 3 3	4 4 5
85	92·20	92·25	92·30	92·36	92·41	92·47	92·52	92·57	92·63	92·68	1 1 2	2 3 3	4 4 5
86	92·74	92·79	92·84	92·90	92·95	93·01	93·06	93·11	93·17	93·22	1 1 2	2 3 3	4 4 5
87	93·27	93·33	93·38	93·43	93·49	93·54	93·59	93·65	93·70	93·75	1 1 2	2 3 3	4 4 5
88	93·81	93·86	93·91	93·97	94·02	94·07	94·13	94·18	94·23	94·29	1 1 2	2 3 3	4 4 5
89	94·34	94·39	94·44	94·50	94·55	94·60	94·66	94·71	94·76	94·82	1 1 2	2 3 3	4 4 5
90	94·87	94·92	94·97	95·03	95·08	95·13	95·18	95·24	95·29	95·34	1 1 2	2 3 3	4 4 5
91	95·39	95·45	95·50	95·55	95·60	95·66	95·71	95·76	95·81	95·86	1 1 2	2 3 3	4 4 5
92	95·92	95·97	96·02	96·07	96·12	96·18	96·23	96·28	96·33	96·38	1 1 2	2 3 3	4 4 5
93	96·44	96·49	96·54	96·59	96·64	96·70	96·75	96·80	96·85	96·90	1 1 2	2 3 3	4 4 5
94	96·95	97·01	97·06	97·11	97·16	97·21	97·26	97·31	97·37	97·42	1 1 2	2 3 3	4 4 5
95	97·47	97·52	97·57	97·62	97·67	97·72	97·78	97·83	97·88	97·93	1 1 2	2 3 3	4 4 5
96	97·98	98·03	98·08	98·13	98·18	98·23	98·29	98·34	98·39	98·44	1 1 2	2 3 3	4 4 5
97	98·49	98·54	98·59	98·64	98·69	98·74	98·79	98·84	98·89	98·94	1 1 2	2 3 3	4 4 5
98	98·99	99·05	99·10	99·15	99·20	99·25	99·30	99·35	99·40	99·45	0 1 1	2 2 3	3 4 4
99	99·50	99·55	99·60	99·65	99·70	99·75	99·80	99·85	99·90	99·95	0 1 1	2 2 3	3 4 4

D 2

	0	1	2	3	4	5	6	7	8	9	1	2	3	4	5	6	7	8	9
10	0·0010000	9901	9804	9709	9615	9524	9434	9346	9259	9174	9	18	27	36	45	55	64	73	82
11	0·0009091	9009	8929	8850	8772	8696	8621	8547	8475	8403	8	15	23	30	38	45	53	61	68
12	0·0008333	8264	8197	8130	8065	8000	7937	7874	7813	7752	6	13	19	26	32	38	45	51	58
13	0·0007692	7634	7576	7519	7463	7407	7353	7299	7246	7194	5	11	16	22	27	33	38	44	49
14	0·0007143	7092	7042	6993	6944	6897	6849	6803	6757	6711	5	10	14	19	24	29	33	38	43
15	0·0006667	6623	6579	6536	6494	6452	6410	6369	6329	6289	4	8	13	17	21	25	29	33	38
16	0·0006250	6211	6173	6135	6098	6061	6024	5988	5952	5917	4	7	11	15	18	22	26	29	33
17	0·0005882	5848	5814	5780	5747	5714	5682	5650	5618	5587	3	6	10	13	16	20	23	26	29
18	0·0005556	5525	5495	5464	5435	5405	5376	5348	5319	5291	3	6	9	12	15	17	20	23	26
19	0·0005263	5236	5208	5181	5155	5128	5102	5076	5051	5025	3	5	8	11	13	16	18	21	24
20	0·0005000	4975	4950	4926	4902	4878	4854	4831	4808	4785	2	5	7	10	12	14	17	19	21
21	0·0004762	4739	4717	4695	4673	4651	4630	4608	4587	4566	2	4	7	9	11	13	15	17	20
22	0·0004545	4525	4505	4484	4464	4444	4425	4405	4386	4367	2	4	6	8	10	12	14	16	18
23	0·0004348	4329	4310	4292	4274	4255	4237	4219	4202	4184	2	4	5	7	9	11	13	14	16
24	0·0004167	4149	4132	4115	4098	4082	4065	4049	4032	4016	2	3	5	7	8	10	12	13	15
25	0·0004000	3984	3968	3953	3937	3922	3906	3891	3876	3861	2	3	5	6	8	9	11	12	14
26	0·0003846	3831	3817	3802	3788	3774	3759	3745	3731	3717	1	3	4	6	7	8	10	11	13
27	0·0003704	3690	3676	3663	3650	3636	3623	3610	3597	3584	1	3	4	5	7	8	9	11	12
28	0·0003571	3559	3546	3534	3521	3509	3497	3484	3472	3460	1	2	4	5	6	7	9	10	11
29	0·0003448	3436	3425	3413	3401	3390	3378	3367	3356	3344	1	2	3	5	6	7	8	9	10
30	0·0003333	3322	3311	3300	3289	3279	3268	3257	3247	3236	1	2	3	4	5	6	7	9	10
31	0·0003226	3215	3205	3195	3185	3175	3165	3155	3145	3135	1	2	3	4	5	6	7	8	9
32	0·0003125	3115	3106	3096	3086	3077	3067	3058	3049	3040	1	2	3	4	5	6	7	8	9
33	0·0003030	3021	3012	3003	2994	2985	2976	2967	2959	2950	1	2	3	4	4	5	6	7	8
34	0·0002941	2933	2924	2915	2907	2899	2890	2882	2874	2865	1	2	3	3	4	5	6	7	8
35	0·0002857	2849	2841	2833	2825	2817	2809	2801	2793	2786	1	2	2	3	4	5	6	6	7
36	0·0002778	2770	2762	2755	2747	2740	2732	2725	2717	2710	1	2	2	3	4	5	5	6	7
37	0·0002703	2695	2688	2681	2674	2667	2660	2653	2646	2639	1	1	2	3	4	4	5	6	6
38	0·0002632	2625	2618	2611	2604	2597	2591	2584	2577	2571	1	1	2	3	3	4	5	5	6
39	0·0002564	2558	2551	2545	2538	2532	2525	2519	2513	2506	1	1	2	3	3	4	4	5	6
40	0·0002500	2494	2488	2481	2475	2469	2463	2457	2451	2445	1	1	2	2	3	4	4	5	5
41	0·0002439	2433	2427	2421	2415	2410	2404	2398	2392	2387	1	1	2	2	3	3	4	5	5
42	0·0002381	2375	2370	2364	2358	2353	2347	2342	2336	2331	1	1	2	2	3	3	4	4	5
43	0·0002326	2320	2315	2309	2304	2299	2294	2288	2283	2278	1	1	2	2	3	3	4	4	5
44	0·0002273	2268	2262	2257	2252	2247	2242	2237	2232	2227	1	1	2	2	3	3	4	4	5
45	0·0002222	2217	2212	2208	2203	2198	2193	2188	2183	2179	0	1	1	2	2	3	3	4	4
46	0·0002174	2169	2165	2160	2155	2151	2146	2141	2137	2132	0	1	1	2	2	3	3	4	4
47	0·0002128	2123	2119	2114	2110	2105	2101	2096	2092	2088	0	1	1	2	2	3	3	4	4
48	0·0002083	2079	2075	2070	2066	2062	2058	2053	2049	2045	0	1	1	2	2	3	3	3	4
49	0·0002041	2037	2033	2028	2024	2020	2016	2012	2008	2004	0	1	1	2	2	2	3	3	4
50	0·0002000	1996	1992	1988	1984	1980	1976	1972	1969	1965	0	1	1	2	2	2	3	3	4
51	0·0001961	1957	1953	1949	1946	1942	1938	1934	1931	1927	0	1	1	2	2	2	3	3	3
52	0·0001923	1919	1916	1912	1908	1905	1901	1898	1894	1890	0	1	1	1	2	2	3	3	3
53	0·0001887	1883	1880	1876	1873	1869	1866	1862	1859	1855	0	1	1	1	2	2	2	3	3
54	0·0001852	1848	1845	1842	1838	1835	1832	1828	1825	1821	0	1	1	1	2	2	2	3	3

N.B.—Three zeros follow the decimal point in the reciprocal of any four figure whole number except the number 1000.

NOTE.—Numbers in difference columns to be subtracted, not added.—*See Rules.*

	0	1	2	3	4	5	6	7	8	9	1 2 3	4 5 6	7 8 9
55	0·0001818	1815	1812	1808	1805	1802	1799	1795	1792	1789	0 1 1	1 2 2	2 3 3
56	0·0001786	1783	1779	1776	1773	1770	1767	1764	1761	1757	0 1 1	1 2 2	2 3 3
57	0·0001754	1751	1748	1745	1742	1739	1736	1733	1730	1727	0 1 1	1 2 2	2 2 3
58	0·0001724	1721	1718	1715	1712	1709	1706	1704	1701	1698	0 1 1	1 1 2	2 2 3
59	0·0001695	1692	1689	1686	1684	1681	1678	1675	1672	1669	0 1 1	1 1 2	2 2 3
60	0·0001667	1664	1661	1658	1656	1653	1650	1647	1645	1642	0 1 1	1 1 2	2 2 3
61	0·0001639	1637	1634	1631	1629	1626	1623	1621	1618	1616	0 1 1	1 1 2	2 2 2
62	0·0001613	1610	1608	1605	1603	1600	1597	1595	1592	1590	0 1 1	1 1 2	2 2 2
63	0·0001587	1585	1582	1580	1577	1575	1572	1570	1567	1565	0 0 1	1 1 1	2 2 2
64	0·0001563	1560	1558	1555	1553	1550	1548	1546	1543	1541	0 0 1	1 1 1	2 2 2
65	0·0001538	1536	1534	1531	1529	1527	1524	1522	1520	1517	0 0 1	1 1 1	2 2 2
66	0·0001515	1513	1511	1508	1506	1504	1502	1499	1497	1495	0 0 1	1 1 1	2 2 2
67	0·0001493	1490	1488	1486	1484	1481	1479	1477	1475	1473	0 0 1	1 1 1	2 2 2
68	0·0001471	1468	1466	1464	1462	1460	1458	1456	1453	1451	0 0 1	1 1 1	2 2 2
69	0·0001449	1447	1445	1443	1441	1439	1437	1435	1433	1431	0 0 1	1 1 1	2 2 2
70	0·0001429	1427	1425	1422	1420	1418	1416	1414	1412	1410	0 0 1	1 1 1	1 2 2
71	0·0001408	1406	1404	1403	1401	1399	1397	1395	1393	1391	0 0 1	1 1 1	1 2 2
72	0·0001389	1387	1385	1383	1381	1379	1377	1376	1374	1372	0 0 1	1 1 1	1 2 2
73	0·0001370	1368	1366	1364	1362	1361	1359	1357	1355	1353	0 0 1	1 1 1	1 2 2
74	0·0001351	1350	1348	1346	1344	1342	1340	1339	1337	1335	0 0 1	1 1 1	1 1 2
75	0·0001333	1332	1330	1328	1326	1325	1323	1321	1319	1318	0 0 1	1 1 1	1 1 2
76	0·0001316	1314	1312	1311	1309	1307	1305	1304	1302	1300	0 0 1	1 1 1	1 1 2
77	0·0001299	1297	1295	1294	1292	1290	1289	1287	1285	1284	0 0 0	1 1 1	1 1 1
78	0·0001282	1280	1279	1277	1276	1274	1272	1271	1269	1267	0 0 0	1 1 1	1 1 1
79	0·0001266	1264	1263	1261	1259	1258	1256	1255	1253	1252	0 0 0	1 1 1	1 1 1
80	0·0001250	1248	1247	1245	1244	1242	1241	1239	1238	1236	0 0 0	1 1 1	1 1 1
81	0·0001235	1233	1232	1230	1229	1227	1225	1224	1222	1221	0 0 0	1 1 1	1 1 1
82	0·0001220	1218	1217	1215	1214	1212	1211	1209	1208	1206	0 0 0	1 1 1	1 1 1
83	0·0001205	1203	1202	1200	1199	1198	1196	1195	1193	1192	0 0 0	1 1 1	1 1 1
84	0·0001190	1189	1188	1186	1185	1183	1182	1181	1179	1178	0 0 0	1 1 1	1 1 1
85	0·0001176	1175	1174	1172	1171	1170	1168	1167	1166	1164	0 0 0	1 1 1	1 1 1
86	0·0001163	1161	1160	1159	1157	1156	1155	1153	1152	1151	0 0 0	1 1 1	1 1 1
87	0·0001149	1148	1147	1145	1144	1143	1142	1140	1139	1138	0 0 0	1 1 1	1 1 1
88	0·0001136	1135	1134	1133	1131	1130	1129	1127	1126	1125	0 0 0	1 1 1	1 1 1
89	0·0001124	1122	1121	1120	1119	1117	1116	1115	1114	1112	0 0 0	1 1 1	1 1 1
90	0·0001111	1110	1109	1107	1106	1105	1104	1103	1101	1100	0 0 0	1 1 1	1 1 1
91	0·0001099	1098	1096	1095	1094	1093	1092	1091	1089	1088	0 0 0	0 1 1	1 1 1
92	0·0001087	1086	1085	1083	1082	1081	1080	1079	1078	1076	0 0 0	0 1 1	1 1 1
93	0·0001075	1074	1073	1072	1071	1070	1068	1067	1066	1065	0 0 0	0 1 1	1 1 1
94	0·0001064	1063	1062	1060	1059	1058	1057	1056	1055	1054	0 0 0	0 1 1	1 1 1
95	0·0001053	1052	1050	1049	1048	1047	1046	1045	1044	1043	0 0 0	0 1 1	1 1 1
96	0·0001042	1041	1040	1038	1037	1036	1035	1034	1033	1032	0 0 0	0 1 1	1 1 1
97	0·0001031	1030	1029	1028	1027	1026	1025	1024	1022	1021	0 0 0	0 1 1	1 1 1
98	0·0001020	1019	1018	1017	1016	1015	1014	1013	1012	1011	0 0 0	0 1 1	1 1 1
99	0·0001010	1009	1008	1007	1006	1005	1004	1003	1002	1001	0 0 0	0 0 1	1 1 1

N.B.—Three zeros follow the decimal point in the reciprocal of any four figure whole number except the number 1000.

Note.—Numbers in difference columns to be subtracted, not added.—*See Rules.*

	0	1	2	3	4	5	6	7	8	9	1	2	3	4	5	6	7	8	9
1·0	0·0000	0100	0198	0296	0392	0488	0583	0677	0770	0862	10	19	29	38	48	57	67	76	86
1·1	0·0953	1044	1133	1222	1310	1398	1484	1570	1655	1740	9	17	26	35	44	52	61	70	78
1·2	0·1823	1906	1989	2070	2151	2231	2311	2390	2469	2546	8	16	24	32	40	48	56	64	72
1·3	0·2624	2700	2776	2852	2927	3001	3075	3148	3221	3293	7	15	22	30	37	44	52	59	67
1·4	0·3365	3436	3507	3577	3646	3716	3784	3853	3920	3988	7	14	21	28	35	41	48	55	62
1·5	0·4055	4121	4187	4253	4318	4383	4447	4511	4574	4637	6	13	19	26	32	39	45	52	58
1·6	0·4700	4762	4824	4886	4947	5008	5068	5128	5188	5247	6	12	18	24	30	36	42	48	55
1·7	0·5306	5365	5423	5481	5539	5596	5653	5710	5766	5822	6	11	17	23	29	34	40	46	51
1·8	0·5878	5933	5988	6043	6098	6152	6206	6259	6313	6366	5	11	16	22	27	32	38	43	49
1·9	0·6419	6471	6523	6575	6627	6678	6729	6780	6831	6881	5	10	15	20	26	31	36	41	46
2·0	0·6931	6981	7031	7080	7129	7178	7227	7275	7324	7372	5	10	15	20	24	29	34	39	44
2·1	0·7419	7467	7514	7561	7608	7655	7701	7747	7793	7839	5	9	14	19	23	28	33	37	42
2·2	0·7885	7930	7975	8020	8065	8109	8154	8198	8242	8286	4	9	13	18	22	27	31	36	40
2·3	0·8329	8372	8416	8459	8502	8544	8587	8629	8671	8713	4	9	13	17	21	26	30	34	38
2·4	0·8755	8796	8838	8879	8920	8961	9002	9042	9083	9123	4	8	12	16	20	24	29	33	37
2·5	0·9163	9203	9243	9282	9322	9361	9400	9439	9478	9517	4	8	12	16	20	24	27	31	35
2·6	0·9555	9594	9632	9670	9708	9746	9783	9821	9858	9895	4	8	11	15	19	23	26	30	34
2·7	0·9933	9969	$\overline{0}$006	$\overline{0}$043	$\overline{0}$080	$\overline{0}$116	$\overline{0}$152	$\overline{0}$188	$\overline{0}$225	$\overline{0}$260	4	7	11	15	18	22	25	29	33
2·8	1·0296	0332	0367	0403	0438	0473	0508	0543	0578	0613	4	7	11	14	18	21	25	28	32
2·9	1·0647	0682	0716	0750	0784	0818	0852	0886	0919	0953	3	7	10	14	17	20	24	27	31
3·0	1·0986	1019	1053	1086	1119	1151	1184	1217	1249	1282	3	7	10	13	16	20	23	26	30
3·1	1·1314	1346	1378	1410	1442	1474	1506	1537	1569	1600	3	6	10	13	16	19	22	25	29
3·2	1·1632	1663	1694	1725	1756	1787	1817	1848	1878	1909	3	6	9	12	15	18	21	25	28
3·3	1·1939	1969	2000	2030	2060	2090	2119	2149	2179	2208	3	6	9	12	15	18	21	24	27
3·4	1·2238	2267	2296	2326	2355	2384	2413	2442	2470	2499	3	6	9	12	15	17	20	23	26
3·5	1·2528	2556	2585	2613	2641	2669	2698	2726	2754	2782	3	6	8	11	14	17	20	22	25
3·6	1·2809	2837	2865	2892	2920	2947	2975	3002	3029	3056	3	5	8	11	14	16	19	22	25
3·7	1·3083	3110	3137	3164	3191	3218	3244	3271	3297	3324	3	5	8	11	13	16	19	21	24
3·8	1·3350	3376	3403	3429	3455	3481	3507	3533	3558	3584	3	5	8	10	13	16	18	21	23
3·9	1·3610	3635	3661	3686	3712	3737	3762	3788	3813	3838	3	5	8	10	13	15	18	20	23
4·0	1·3863	3888	3913	3938	3962	3987	4012	4036	4061	4085	2	5	7	10	12	15	17	20	22
4·1	1·4110	4134	4159	4183	4207	4231	4255	4279	4303	4327	2	5	7	10	12	14	17	19	22
4·2	1·4351	4375	4398	4422	4446	4469	4493	4516	4540	4563	2	5	7	9	12	14	16	19	21
4·3	1·4586	4609	4633	4656	4679	4702	4725	4748	4770	4793	2	5	7	9	12	14	16	18	21
4·4	1·4816	4839	4861	4884	4907	4929	4951	4974	4996	5019	2	5	7	9	11	14	16	18	20
4·5	1·5041	5063	5085	5107	5129	5151	5173	5195	5217	5239	2	4	7	9	11	13	15	18	20
4·6	1·5261	5282	5304	5326	5347	5369	5390	5412	5433	5454	2	4	6	9	11	13	15	17	19
4·7	1·5476	5497	5518	5539	5560	5581	5602	5623	5644	5665	2	4	6	8	11	13	15	17	19
4·8	1·5686	5707	5728	5748	5769	5790	5810	5831	5851	5872	2	4	6	8	10	12	14	16	19
4·9	1·5892	5913	5933	5953	5974	5994	6014	6034	6054	6074	2	4	6	8	10	12	14	16	18
5·0	1·6094	6114	6134	6154	6174	6194	6214	6233	6253	6273	2	4	6	8	10	12	14	16	18
5·1	1·6292	6312	6332	6351	6371	6390	6409	6429	6448	6467	2	4	6	8	10	12	14	16	18
5·2	1·6487	6506	6525	6544	6563	6582	6601	6620	6639	6658	2	4	6	8	10	11	13	15	17
5·3	1·6677	6696	6715	6734	6752	6771	6790	6808	6827	6845	2	4	6	7	9	11	13	15	17
5·4	1·6864	6882	6901	6919	6938	6956	6974	6993	7011	7029	2	4	5	7	9	11	13	15	16

TABLE OF NEPERIAN LOGARITHMS OF 10^{+n}

n	1	2	3	4	5	6	7	8	9
$\log_e 10^n$	2·3026	4·6052	6·9078	9·2103	11·5129	13·8155	16·1181	18·4207	20·7233

	0	1	2	3	4	5	6	7	8	9	1	2	3	4	5	6	7	8	9
5·5	1·7047	7066	7084	7102	7120	7138	7156	7174	7192	7210	2	4	5	7	9	11	13	14	16
5·6	1·7228	7246	7263	7281	7299	7317	7334	7352	7370	7387	2	4	5	7	9	11	12	14	16
5·7	1·7405	7422	7440	7457	7475	7492	7509	7527	7544	7561	2	3	5	7	9	10	12	14	16
5·8	1·7579	7596	7613	7630	7647	7664	7681	7699	7716	7733	2	3	5	7	9	10	12	14	15
5·9	1·7750	7766	7783	7800	7817	7834	7851	7867	7884	7901	2	3	5	7	8	10	12	13	15
6·0	1·7918	7934	7951	7967	7984	8001	8017	8034	8050	8066	2	3	5	7	8	10	12	13	15
6·1	1·8083	8099	8116	8132	8148	8165	8181	8197	8213	8229	2	3	5	6	8	10	11	13	15
6·2	1·8245	8262	8278	8294	8310	8326	8342	8358	8374	8390	2	3	5	6	8	10	11	13	14
6·3	1·8405	8421	8437	8453	8469	8485	8500	8516	8532	8547	2	3	5	6	8	9	11	13	14
6·4	1·8563	8579	8594	8610	8625	8641	8656	8672	8687	8703	2	3	5	6	8	9	11	12	14
6·5	1·8718	8733	8749	8764	8779	8795	8810	8825	8840	8856	2	3	5	6	8	9	11	12	14
6·6	1·8871	8886	8901	8916	8931	8946	8961	8976	8991	9006	2	3	5	6	8	9	11	12	14
6·7	1·9021	9036	9051	9066	9081	9095	9110	9125	9140	9155	1	3	4	6	7	9	10	12	13
6·8	1·9169	9184	9199	9213	9228	9242	9257	9272	9286	9301	1	3	4	6	7	9	10	12	13
6·9	1·9315	9330	9344	9359	9373	9387	9402	9416	9430	9445	1	3	4	6	7	9	10	12	13
7·0	1·9459	9473	9488	9502	9516	9530	9544	9559	9573	9587	1	3	4	6	7	9	10	11	13
7·1	1·9601	9615	9629	9643	9657	9671	9685	9699	9713	9727	1	3	4	6	7	8	10	11	13
7·2	1·9741	9755	9769	9782	9796	9810	9824	9838	9851	9865	1	3	4	6	7	8	10	11	12
7·3	1·9879	9892	9906	9920	9933	9947	9961	9974	9988	0001	1	3	4	5	7	8	10	11	12
7·4	2·0015	0028	0042	0055	0069	0082	0096	0109	0122	0136	1	3	4	5	7	8	9	11	12
7·5	2·0149	0162	0176	0189	0202	0215	0229	0242	0255	0268	1	3	4	5	7	8	9	11	12
7·6	2·0281	0295	0308	0321	0334	0347	0360	0373	0386	0399	1	3	4	5	7	8	9	10	12
7·7	2·0412	0425	0438	0451	0464	0477	0490	0503	0516	0528	1	3	4	5	6	8	9	10	12
7·8	2·0541	0554	0567	0580	0592	0605	0618	0631	0643	0656	1	3	4	5	6	8	9	10	12
7·9	2·0669	0681	0694	0707	0719	0732	0744	0757	0769	0782	1	3	4	5	6	8	9	10	11
8·0	2·0794	0807	0819	0832	0844	0857	0869	0882	0894	0906	1	3	4	5	6	8	9	10·11	
8·1	2·0919	0931	0943	0956	0968	0980	0992	1005	1017	1029	1	2	4	5	6	7	9	10	11
8·2	2·1041	1054	1066	1078	1090	1102	1114	1126	1138	1150	1	2	4	5	6	7	9	10	11
8·3	2·1163	1175	1187	1199	1211	1223	1235	1247	1258	1270	1	2	4	5	6	7	8	10	11
8·4	2·1282	1294	1306	1318	1330	1342	1353	1365	1377	1389	1	2	4	5	6	7	8	10	11
8·5	2·1401	1412	1424	1436	1448	1459	1471	1483	1494	1506	1	2	4	5	6	7	8	9	11
8·6	2·1518	1529	1541	1552	1564	1576	1587	1599	1610	1622	1	2	3	5	6	7	8	9	10
8·7	2·1633	1645	1656	1668	1679	1691	1702	1713	1725	1736	1	2	3	5	6	7	8	9	10
8·8	2·1748	1759	1770	1782	1793	1804	1815	1827	1838	1849	1	2	3	5	6	7	8	9	10
8·9	2·1861	1872	1883	1894	1905	1917	1928	1939	1950	1961	1	2	3	4	6	7	8	9	10
9·0	2·1972	1983	1994	2006	2017	2028	2039	2050	2061	2072	1	2	3	4	6	7	8	9	10
9·1	2·2083	2094	2105	2116	2127	2138	2148	2159	2170	2181	1	2	3	4	5	7	8	9	10
9·2	2·2192	2203	2214	2225	2235	2246	2257	2268	2279	2289	1	2	3	4	5	6	8	9	10
9·3	2·2300	2311	2322	2332	2343	2354	2364	2375	2386	2396	1	2	3	4	5	6	7	9	10
9·4	2·2407	2418	2428	2439	2450	2460	2471	2481	2492	2502	1	2	3	4	5	6	7	8	10
9·5	2·2513	2523	2534	2544	2555	2565	2576	2586	2597	2607	1	2	3	4	5	6	7	8	9
9·6	2·2618	2628	2638	2649	2659	2670	2680	2690	2701	2711	1	2	3	4	5	6	7	8	9
9·7	2·2721	2732	2742	2752	2762	2773	2783	2793	2803	2814	1	2	3	4	5	6	7	8	9
9·8	2·2824	2834	2844	2854	2865	2875	2885	2895	2905	2915	1	2	3	4	5	6	7	8	9
9·9	2·2925	2935	2946	2956	2966	2976	2986	2996	3006	3016	1	2	3	4	5	6	7	8	9

TABLE OF NEPERIAN LOGARITHMS OF 10^{-n}.

n	1	2	3	4	5	6	7	8	9
$\log_e 10^{-n}$	$\bar{3}\cdot6974$	$\bar{5}\cdot3948$	$\bar{7}\cdot0922$	$\overline{10}\cdot7897$	$\overline{12}\cdot4871$	$\overline{14}\cdot1845$	$\overline{17}\cdot8819$	$\overline{19}\cdot5793$	$\overline{21}\cdot2767$

Table of Powers of ϵ;—ϵ being the Base of the Hyperbolic Logarithms.

Power	Value	Power	Value	Power	Value
ϵ^{1}	2·7183	$\epsilon^{\frac{1}{2}}$	1·6487	$\epsilon^{\frac{1}{2}}$	1·6487
ϵ^{2}	7·3891	$\epsilon^{\frac{3}{2}}$	4·4816	$\epsilon^{\frac{1}{3}}$	1·3956
ϵ^{3}	20·086	$\epsilon^{\frac{5}{2}}$	12·182	$\epsilon^{\frac{1}{4}}$	1·2840
ϵ^{4}	54·598	$\epsilon^{\frac{7}{2}}$	33·114	$\epsilon^{\frac{1}{5}}$	1·2214
ϵ^{5}	148·41	$\epsilon^{\frac{1}{4}}$	1·2840	$\epsilon^{\frac{1}{6}}$	1·1814
ϵ^{6}	403·43	$\epsilon^{\frac{1}{8}}$	1·1331	$\epsilon^{\frac{1}{7}}$	1·1536
ϵ^{7}	1096·6	$\epsilon^{\frac{1}{16}}$	1·0645	$\epsilon^{\frac{1}{8}}$	1·1331
ϵ^{8}	2981·0	$\epsilon^{\frac{1}{32}}$	1·0317	$\epsilon^{\frac{1}{9}}$	1·1175
ϵ^{9}	8103·1			$\epsilon^{\frac{1}{10}}$	1·1052
ϵ^{10}	22026·				

Power	Value	Power	Value	Power	Value
ϵ^{-1}	0·3679	$\epsilon^{-\frac{1}{2}}$	0·6065	$\epsilon^{-\frac{1}{2}}$	0·6065
ϵ^{-2}	0·1353	$\epsilon^{-\frac{3}{2}}$	0·2231	$\epsilon^{-\frac{1}{3}}$	0·7165
ϵ^{-3}	0·04979	$\epsilon^{-\frac{5}{2}}$	0·0821	$\epsilon^{-\frac{1}{4}}$	0·7788
ϵ^{-4}	0·01832	$\epsilon^{-\frac{7}{2}}$	0·0302	$\epsilon^{-\frac{1}{5}}$	0·8187
ϵ^{-5}	0·006738	$\epsilon^{-\frac{1}{4}}$	0·7788	$\epsilon^{-\frac{1}{6}}$	0·8465
ϵ^{-6}	0·002479	$\epsilon^{-\frac{1}{8}}$	0·8825	$\epsilon^{-\frac{1}{7}}$	0·8669
ϵ^{-7}	0·0009119	$\epsilon^{-\frac{1}{16}}$	0·9394	$\epsilon^{-\frac{1}{8}}$	0·8825
ϵ^{-8}	0·0003355	$\epsilon^{-\frac{1}{32}}$	0·9692	$\epsilon^{-\frac{1}{9}}$	0·8948
ϵ^{-9}	0·0001234			$\epsilon^{-\frac{1}{10}}$	0·9048
ϵ^{-10}	0·0000454				

Power	Value	Power	Value	Power	Value
ϵ^{π}	23·1407	$\epsilon^{\frac{\pi}{2}}$	4·8105	$\epsilon^{\frac{\pi}{4}}$	2·1933
$\epsilon^{2\pi}$	535·491	$\epsilon^{\frac{3\pi}{2}}$	111·32	$\epsilon^{\frac{3\pi}{4}}$	10·557
$\epsilon^{3\pi}$	12391·7	$\epsilon^{\frac{5\pi}{2}}$	2576·0	$\epsilon^{\frac{5\pi}{4}}$	50·754
$\epsilon^{4\pi}$	286752·	$\epsilon^{\frac{7\pi}{2}}$	59609·6	$\epsilon^{\frac{7\pi}{4}}$	244·15
$\epsilon^{5\pi}$	6635627	$\epsilon^{\frac{9\pi}{2}}$	1379406	$\epsilon^{\frac{9\pi}{4}}$	1174·48

Power	Value	Power	Value	Power	Value
$\epsilon^{-\pi}$	43214 × 10⁻⁶	$\epsilon^{-\frac{\pi}{2}}$	20788 × 10⁻⁵	$\epsilon^{-\frac{\pi}{4}}$	4559 × 10⁻⁴
$\epsilon^{-2\pi}$	1867 × ,,	$\epsilon^{-\frac{3\pi}{2}}$	898·33 × ,,	$\epsilon^{-\frac{3\pi}{4}}$	947·8 × ,,
$\epsilon^{-3\pi}$	80·699 × ,,	$\epsilon^{-\frac{5\pi}{2}}$	38·820 × ,,	$\epsilon^{-\frac{5\pi}{4}}$	197·0 × ,,
$\epsilon^{-4\pi}$	3·487 × ,,	$\epsilon^{-\frac{7\pi}{2}}$	1·6776 × ,,	$\epsilon^{-\frac{7\pi}{4}}$	40·96 × ,,
$\epsilon^{-5\pi}$	·1507 × ,,	$\epsilon^{-\frac{9\pi}{2}}$	·0725 × ,,	$\epsilon^{-\frac{9\pi}{4}}$	8·514 × ,,

USEFUL FORMULAS AND NUMBERS.

Binomial Theorem.

$$(1 \pm e)^n = 1 \pm ne + \frac{n.\overline{n-1}}{1.2}e^2 \pm \frac{n.\overline{n-1}.\overline{n-2}}{1.2.3}e^3 + \&c.$$

Hence, when ne is so small that its square and higher powers may be neglected, $(1 \pm e)^n \fallingdotseq 1 \pm ne$.

Examples—

$e = \text{·}01$; $(1 + \text{·}01)^2 \fallingdotseq 1\text{·}02$; $(1 + \text{·}01)^{\frac{1}{2}} \fallingdotseq 1\text{·}005$; $(1 + \text{·}01)^{-\frac{1}{3}} \fallingdotseq 0\text{·}9967$.

Barometric Formula.—Let P and p be the atmospheric pressures observed by the barometer at the lower and upper stations respectively ; and let T and t be the respective atmospheric temperatures on the Fahrenheit scale ; then H, being the difference of levels in feet,

$$H = 60360 \left\{\log P - \log p\right\} \left(1 + \frac{T+t-64}{986}\right).$$

Base of Hyperbolic or Neperian Logarithms, ... $\epsilon = 2\text{·}71828$.
To convert common into hyperbolic logarithms, multiply by . . . . 2·30258.
To convert hyperbolic into common logarithms, multiply by . . . . 0·43429.
Ratio of circumference of circle to diameter, ... $\pi = 3\text{·}14159$.
Number of degrees in one radian (the unit angle, which is the angle subtended by arc equal to radius), $57°\text{·}2958 = 57°\ 17'\ 45'' = 206265''$.

		Logarithm.
$\pi = 3\text{·}14159$	0·49715
$\epsilon = 2\text{·}71828$	0·43429

Metre in inches,	.	39·37043	Cubic inch of distilled	
Foot in centimetres,	.	30·4797	water at 4°C, . .	252·89 grains.
Mile in kilometres,	.	1·6093	Cubic foot of water at	
Gramme in grains,	.	15·43235	4°C, ...	62·43 lbs.
Pound in grammes,	.	453·593	Cubic inch of mercury	
Kilogramme in pounds,	.	2·2046	at 0°C, . . .	3439 grains.
British ton in French tons			Do. do.,	·4913 lbs.
of 1000 kilos.,	.	1·016	Litre of dry air at 0°C,	
Litre in cubic inches,	.	61·0253	760ᵐ·ᵐ· pressure, .	1·2932 grms.
Cubic inch in cubic centimetres,	16·3866	Cubic foot ,, ,,	565·1 grains.
Cubic foot in cubic centimetres,	28316·0	Density of mercury, .	13·596.

1 centim. gramme . . . . 981 ergs.
1 metre kilogramme . . . . 9.81×10^7.
1 ft. lb. . . . . 13.56×10^8 ergs.
1 ft. poundal (independent of g.) . . . . . 421390 ergs.
1 joule (1 watt for 1 second) . . . . . . 10^7 ergs.
1 horse power. . . . . 7.46×10^9 ergs per sec.
1 watt (rate of working of 1 volt through 1 ohm, or of 1 volt carrying 1 ampere)
  $= 10^7$ ergs per second.

Earth's mean radius, 6.371×10^8 centims
Earth's mean radius
  (approx.), . . $.21 \times 10^6$ feet.
Mass of earth, as-
  suming 5.67 as
  mean density, . 6.14×10^{27} grammes
Earth's mass (ap-
  proximately), . 13.5×10^{24} lbs. $=$
  6×10^{21} tons.

Mass of moon, . 1/81.5 of earth's mass
Distance of moon
  from earth, . 3.8×10^{10} centims.
Sun's radius, . 697×10^8 centims.
Mass of sun, . 324000 earth's mass.
Distance of sun
  from earth, . 1.498×10^{13} centims.
Distance of sun
  from earth, . 93.1×10^6 miles.

Seconds pendulum at
  Greenwich, . . 39.139 inches $=$
    99.414 centims.
Gravity of 1 pound
  at Greenwich, . 32.191 poundals.
Gravity of 1 pound
  mass in lat. 55° 35'
  (approximately that
  of Edinburgh or
  Glasgow), . . 32.2 poundals.
Gravity of 1 gramme
  in same latitude, . 981.424 dynes.

Mass in grammes which
  concentrated at a
  point 1 centimetre
  distant from a point at
  which another equal
  mass is concentrated
  would attract it with a
  force of 1 dyne, 3928 grammes.
Same where the foot,
  pound, and poundal
  are units of length,
  mass, and force, 31,075 lbs.

Height of Homogeneous atmosphere at Greenwich at 0°C, 26,210 ft. $=$
  7.988×10^5 centims.

Newtonian velocity of sound in air at 0°C, . . 918.5 feet per second,
  $= 27996$ centims. per second.

True velocity at t°C $= 33240 \sqrt{1 + .00366\, t}$ centims. per second.

Joule's Equivalent. 777.2 Greenwich foot-pounds of work will raise
  1 lb. pure water from 60° to 61° Fahrenheit.

  This is equivalent to 1399 ft.-lbs. per pound degree centigrade,—or
  41.84×10^6 ergs per gramme degree centigrade,—or 42600 centi-
  metre-grammes per gramme degree.

Latent heat of water, 79.25. Latent heat of steam at 100°, 537

Specific heat of air pressure constant, 0.237 ;

$$\frac{\text{sp. heat of air pres. const.}}{\text{sp. heat of air vol. const.}} = 1.4.$$

1 litre of hydrogen at 0°C and 760 mm. pressure weighs 0.0896 gm.

Density of hydrogen compared with air $= 0.0693 = \dfrac{1}{14.43}$.

Conductivity of heat. Quantity, in gramme-water-centigrade units; conducted per second; per square centimetre of area; per degree, per centimetre of thickness, of difference of temperature of two sides of plate.

Copper,.............................'996
Iron,.................................'15 to '19
Stone,...............................'01 to '005

Velocity of light in vacuum $= 3\cdot004 \times 10^{10}$ centims. per second.

Mean wave length $5\cdot3 \times 10^{-5}$ centim.

One electromagnetic unit $= 3 \times 10^{10}$ electrostatic units of electricity

1 B. A. Unit = '9866 Ohm.   1 Ohm = 1'01358 B. A. Unit.

Resistance of 100 metres of pure annealed round wire, 1 mm. in diameter, at 0°C,

		Copper,	.	.	.	.	2'028 ohms.
,,	,,	Iron,.	.	.	.	.	12'34 ,,
,,	,,	Platinum,	.	.	.	.	11'50 ,,
,,	,,	Platinoid,	.	.	.	.	41'17 ,,

Electro-motive force of 1 Daniell's cell, ... 1'07 volt.

,, ,, 1 Grove's cell, ... 1'95 ,,

,, ,, Standard Clark cell at 15°C, . . 1'435 volt.

,, ,, ,, ,, t°C $1\cdot435[1 - \cdot00077(t-15)]$

One volt through one ohm (1 watt) generates per second $\dfrac{1}{4\cdot184}$ of a thermal unit (gramme-water-centigrade).

Electro-Chemical Decomposition.—

Element.	Atomic Weight.	Chem. Equivalent.	Electrolytic Decomposition, Grammes per second, per ampere.
Hydrogen,	1	1	'00001038
Potassium,	39'03	39'03	'0004051
Sodium,	23'	23'	'0002387
Silver,	107'7	107'7	'001118
Copper,*	63'35	31'68	'0003290
Zinc,	64'88	32'44	'0003367
Lead,	206'4	103'2	'001071
Oxygen,	15'96	7'98	'00008283
Chlorine,	35'37	35'37	'0003671

* For cathode surface of 50 sq. centims. per ampere, the quantity of copper deposited per ampere per second is '0 03287 gms. For increments in cathode surface subtract 1·16th p.c. per 50 sq. centims. The numbers given for silver and copper are the results of direct experiment

GLASGOW: PRINTED AT THE UNIVERSITY PRESS
BY ROBERT MACLEHOSE AND CO. LTD.